T0396186

VOLUME NINETY THREE

CURRENT TOPICS IN MEMBRANES

Vesicle Trafficking in Eukaryotes

CURRENT TOPICS IN MEMBRANES, VOLUME 93

Series Editor

IRENA LEVITAN

Division of Pulmonary and Critical Care Medicine
University of Illinois at Chicago
Chicago, IL, United States

VOLUME NINETY THREE

CURRENT TOPICS IN MEMBRANES

Vesicle Trafficking in Eukaryotes

Edited by

LUCIANA ANDRADE

Department of Morphology
Federal University of Minas Gerais
Belo Horizonte, MG, Brazil

CHRISTOPHER KUSHMERICK

Department of Physiology and Biophysics
Federal University of Minas Gerais
Belo Horizonte, MG, Brazil

Academic Press is an imprint of Elsevier
125 London Wall, London, EC2Y 5AS, United Kingdom
50 Hampshire Street, 5th Floor, Cambridge, MA 02139, United States
525 B Street, Suite 1650, San Diego, CA 92101, United States

First edition 2024

ISBN: 978-0-443-29458-7
ISSN: 1063-5823

For information on all Academic Press publications
visit our website at https://www.elsevier.com/books-and-journals

Publisher: Zoe Kruze
Acquisitions Editor: Leticia Lima
Editorial Project Manager: Sneha Apar
Production Project Manager: Abdulla Sait
Cover Designer: Gopalakrishnan Venkatraman

Typeset by MPS Limited, India

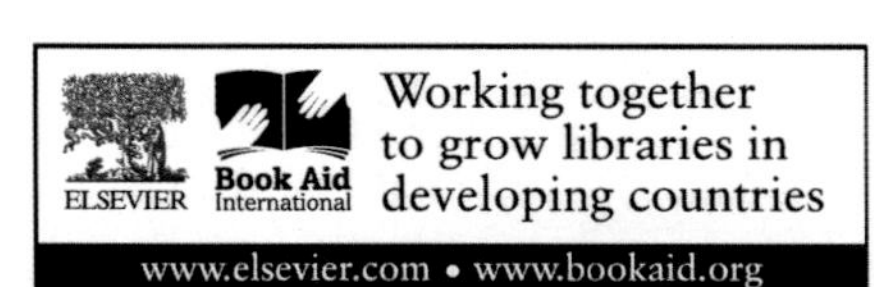

Contents

Contributors

Daniel Adebayo
Department of Biological Sciences, Wayne State University, Detroit, MI, United States

Sumit Bandyopadhyay
Department of Biological Sciences, Wayne State University, Detroit, MI, United States

Assaf Biran
Department of Biomolecular Sciences, Weizmann Institute of Science, Rehovot, Israel

Celso Caruso-Neves
Carlos Chagas Filho Biophysics Institute, Federal University of Rio de Janeiro; Rio de Janeiro Innovation Network in Nanosystems for Health—NanoSAUDE/FAPERJ, Rio de Janeiro, RJ, Brazil; Instituto Nacional de Ciência e Tecnologia em Medicina Regenerativa, INCT-Regenera, Conselho

Tamir Dingjan
Department of Biomolecular Sciences, Weizmann Institute of Science, Rehovot, Israel

Anthony H. Futerman
Department of Biomolecular Sciences, Weizmann Institute of Science, Rehovot, Israel

Hanaa Hariri
Department of Biological Sciences, Wayne State University, Detroit, MI, United States

Kildare Miranda
Laboratório de Ultraestrutura Celular Hertha Meyer, Instituto de Biofísica Carlos Chagas Filho and Centro Nacional de Biologia Estrutural e Bioimagem, Universidade Federal do Rio de Janeiro, Rio de Janeiro, Brazil

Eseiwi Obaseki
Department of Biological Sciences, Wayne State University, Detroit, MI, United States

Laura B.F. Oliveira
Department of Physiology and Biophysics, Institute of Biological Sciences, Federal University of Minas Gerais, Belo Horizonte, MG, Brazil

Diogo B. Peruchetti
Department of Physiology and Biophysics, Institute of Biological Sciences, Federal University of Minas Gerais; Instituto Nacional de Ciência e Tecnologia em Nanobiofarmacêutica, INCT-NANOBiofar, Conselho Nacional de Desenvolvimento Científico e Tecnológico/ MCTI, Belo Horizonte, MG, Brazil

Mariana C. Rodrigues
Department of Physiology and Biophysics, Institute of Biological Sciences, Federal University of Minas Gerais, Belo Horizonte, MG, Brazil

Adyatma Irawan Santosa
Department of Plant Protection, Faculty of Agriculture, Universitas Gadjah Mada, Yogyakarta, Indonesia

Maria Aparecida R. Vieira
Department of Physiology and Biophysics, Institute of Biological Sciences, Federal University of Minas Gerais, Belo Horizonte, MG, Brazil

Camila Wendt
Laboratório de Ultraestrutura Celular Hertha Meyer, Instituto de Biofísica Carlos Chagas Filho and Centro Nacional de Biologia Estrutural e Bioimagem; Laboratório de Biomineralização, Instituto de Ciências Biomédicas, Universidade Federal do Rio de Janeiro, Rio de Janeiro, Brazil

Ali Çelik
Department of Plant Protection, Faculty of Agriculture, Bolu Abant İzzet Baysal University, Bolu, Türkiye

Preface

We organized this volume of *Current Topics in Membranes* to explore aspects of vesicle trafficking in normal cell physiology and disease. In Chapter 1, Rodrigues et al. review receptor-mediated endocytosis throughout the nephron and its role in renal physiology and disease. In Chapter 2, Wendt and Miranda present results obtained from volume EM imaging techniques used to study endocytosis of host hemoglobin by Plasmodium, the parasite that causes malaria. In Chapter 3, Biran, Dingjan and Futerman describe the distribution of enzymes for sphingolipid synthesis in the endoplasmic reticulum and the golgi, and intracellular vesicles that carry intermediates from the former to the latter. In Chapter 4, Çelik and Santosa use the phylogenetic relationship of nucleotide sequences for coat protein to explore the evolution of plant Ilarvirus. In Chapter 5, Bandyopadhyay et al., review the lysosome with a focus on lysosomal interactions with other organelles.

CHAPTER ONE

Receptor-mediated endocytosis in kidney cells during physiological and pathological conditions

Mariana C. Rodrigues[a,1], Laura B.F. Oliveira[a,1], Maria Aparecida R. Vieira[a], Celso Caruso-Neves[b,c,d], and Diogo B. Peruchetti[a,e,*]

[a]Department of Physiology and Biophysics, Institute of Biological Sciences, Federal University of Minas Gerais, Belo Horizonte, MG, Brazil
[b]Carlos Chagas Filho Biophysics Institute, Federal University of Rio de Janeiro, Rio de Janeiro, RJ, Brazil
[c]Rio de Janeiro Innovation Network in Nanosystems for Health—NanoSAUDE/FAPERJ, Rio de Janeiro, RJ, Brazil
[d]Instituto Nacional de Ciência e Tecnologia em Medicina Regenerativa, INCT-Regenera, Conselho Nacional de Desenvolvimento Científico e Tecnológico/MCTI, Rio de Janeiro, RJ, Brazil
[e]Instituto Nacional de Ciência e Tecnologia em Nanobiofarmacêutica, INCT-NANOBiofar, Conselho Nacional de Desenvolvimento Científico e Tecnológico/MCTI, Belo Horizonte, MG, Brazil
*Corresponding author. e-mail address: dperuchetti@ufmg.br

Contents

Abstract

Mammalian cell membranes are very dynamic where they respond to several environmental stimuli by rearranging the membrane composition by basic biological processes, including endocytosis. In this context, receptor-mediated endocytosis, either clathrin-dependent or caveolae-dependent, is involved in different physiological and pathological

[1] MRC and LBFO contributed equally to the development of this work.

Current Topics in Membranes, Volume 93
ISSN 1063-5823, https://doi.org/10.1016/bs.ctm.2024.05.003

conditions. In the last years, an important amount of evidence has been reported that kidney function involves the modulation of different types of endocytosis, including renal protein handling. In addition, the dysfunction of the endocytic machinery is involved with the development of proteinuria as well as glomerular and tubular injuries observed in kidney diseases associated with hypertension, diabetes, and others. In this present review, we will discuss the mechanisms underlying the receptor-mediated endocytosis in different glomerular cells and proximal tubule epithelial cells as well as their modulation by different factors during physiological and pathological conditions. These findings could help to expand the current understanding regarding renal protein handling as well as identify possible new therapeutic targets to halt the progression of kidney disease.

1. Introduction

Among the eukaryotic cells, the mammalian cell membranes are very dynamic, and they also seek to respond appropriately to several environmental stimuli by rearranging the membrane composition through, at least, two basic cellular processes such as endocytosis and exocytosis. As a result, the cells can adapt to many microenvironments observed during physiological and pathological conditions. Among these, endocytosis is the process involving the transport of extracellular materials or cargo into intracellular compartments by a series of pathways followed by the formation of intracellular vesicles (Cooper & McNeil, 2015; McMahon & Boucrot, 2011). A consensus is now developing for five major types of endocytosis: (1) clathrin-mediated endocytosis (clathrin and dynamin-dependent), (2) fast endophilin-mediated endocytosis (a clathrin-independent but dynamin-dependent pathway for rapid ligand-driven endocytosis of specific membrane proteins), (3) clathrin-independent carrier (CLIC)/glycosylphosphatidylinositol-anchored protein-enriched early endocytic compartment (GEEC) endocytosis (clathrin and dynamin independent), (4) macropinocytosis and (5) phagocytosis (Rennick, Johnston, & Parton, 2021). Furthermore, caveolae-dependent endocytosis represents a sixth pathway that also contribute to endocytic uptake. These endocytic pathways are involved in many cellular functions to maintain both cellular and body homeostasis (Cooper & McNeil, 2015; McMahon & Boucrot, 2011). Furthermore, it is known that these endocytic pathways are tightly regulated by various signaling molecules (Gekle, Mildenberger, Freudinger, Schwerdt, & Silbernagl, 1997; Hryciw, Pollock, & Poronnik, 2005; Koral et al., 2014; Peres et al., 2023; Silva-Aguiar et al., 2022). In the last years, there have been a significant number of studies showing that the deregulation of endocytic pathways is correlated to the development and progression of kidney diseases (Peres et al., 2023; Peruchetti et al., 2020, 2021; Silva-Aguiar et al., 2022; Teixeira et al., 2019). In this context, the integration of the current understanding underlying how modulation of endocytosis in renal cells results in kidney pathological processes is a very important issue to be studied.

The kidneys are the main responsible for maintaining the body's homeostasis through the regulation of plasma osmolarity, extracellular volume, and blood pressure as well as acid-base equilibrium. These regulatory processes are efficiently performed by the nephrons (the renal morpho functional units) through three basic functions: (1) glomerular filtration; (2) tubular reabsorption; and/or (3) tubular secretion. Under physiological conditions, the integrity of the glomerular filtration barrier (GFB) provides good restriction for the passage of high-molecular-weight plasma macromolecules. Despite these restrictive properties, it has already been shown that a significant amount of plasma components are filtered through the GFB (Haraldsson, Nyström, & Deen, 2008; Moeller & Tenten, 2013). The classical view indicates that this process occurs due to paracellular pathways. However, evidence showed that the filtration of macromolecules can be more complex than thought. Once filtered, plasma proteins are reabsorbed by proximal tubule epithelial cells (PTECs) by receptor-mediated endocytosis (Abbate, Zoja, & Remuzzi, 2006; Dickson, Wagner, Sandoval, & Molitoris, 2014; Gekle, 2005). In this way, glomerular damage and/or tubular reabsorption dysfunction will impact the development of low- from high-molecular-weight proteinuria as well as the development of kidney disease (Abbate et al., 2006; Dickson et al., 2014). This present review will discuss the different studies showing the mechanisms of receptor-mediated endocytosis in renal cells as well as the regulatory mechanisms underlying this process during physiological and pathophysiological conditions.

2. Glomerular cells and endocytosis

The glomerulus consists of glomerular capillary loops surrounded by the Bowman's capsule which is formed by the parietal epithelial cell (PEC) layer. The glomerulus starts the urine formation process due to glomerular ultrafiltration of plasma: the passage of fluid from the vascular compartment to the urinary space compartment. This mechanism involves two factors: the effective ultrafiltration pressure (a result of the balance among Starling forces) and the ultrafiltration coefficient (also known as Kf). The first one is the driving force for the transport of different compounds and water across the GFB. Now, the second one involves the permeability of GFB determined by the integrity of three layers: (1) fenestrated endothelium from glomerular capillary loops; (2) the glomerular basal membrane; (3) podocytes, forming the internal glomerular epithelial cell layer. In addition, mesangial cells are stromal cells involved in the regulation of glomerular capillary blood flow. Glomerular filtration is known to be a passive process due to the paracellular transport mechanisms. The GFB offers restrictions for the passage of blood cells and the high-molecular-weight macromolecules from plasma. Using different technical approaches, it has been suggested that the amount of plasma proteins in the urinary space varies during physiological and pathological conditions. Interestingly, as showed below, different studies have reported that glomerular cells also can perform endocytosis of plasma proteins to achieve different purposes. To better understand this phenomenon, the potential contribution of protein endocytosis on glomerular endothelial cells (GECs), podocytes, mesangial cells, and PECs for glomerular filtration of plasma proteins will be discussed below.

2.1 Glomerular endothelial cells and receptor-mediated endocytosis

GECs form the first layer of the GFB and are very important in determining the size and charge restriction to the plasma macromolecules (Haraldsson et al., 2008). GECs present variated size fenestrae which already limits the passage of several components from the bloodstream into the glomerular ultrafiltrate (Haraldsson et al., 2008; Moeller & Tenten, 2013). Beyond the fenestrae, the GECs present a complex glycocalyx surface layer that works as an efficient selective molecular filter contributing to the maintenance of the glomerular selective permeability (Haraldsson et al., 2008). Interestingly, GECs possess membrane microdomains (e.g. caveolae) and endocytic vesicles known to be involved in the protein internalization (Moriyama et al., 2015; Moriyama, Karasawa, & Nitta, 2018), which suggests that these cells could present a major role in plasma proteins filtration than previously thought. In this section, it will be discussed the potential contribution of endocytosis in GECs for protein filtration by glomerulus in physiological and pathological conditions.

Moriyama et al. (2015) showed that human renal GECs promote albumin endocytosis. In addition, the authors showed that Alexa Fluor 488-labeled bovine serum albumin (BSA) co-localized with caveolin-1 (Cav-1) but not with clathrin, indicating that this process occurs through the caveolae. Later, it was shown that endocytosis of albumin by GECs involves intracellular transcytosis (Moriyama, Sasaki, Karasawa, Uchida, & Nitta, 2017).

Since caveolae provide a pathway involved in albumin entrance in GECs, it could be important in the pathogenesis of albuminuria. Wu et al. (2016), using the *db/db* mice model, showed that salidroside exerts its proteinuria-alleviating effects by downregulation of Cav-1 phosphorylation and inhibition of albumin transcytosis across GECs. The authors propose that interference with albumin transcytosis across GECs is a novel approach to the treatment of albuminuria in diabetes. However, some questions still need to be addressed: (1) what's the mechanism involved in this process? (2) what are the receptors involved? (3) What are the signaling pathways associated with transcytosis in GECs being modulated during the development of albuminuria? Future studies are important to address these issues.

2.2 Podocytes and receptor-mediated endocytosis

Podocytes are polarized epithelial cells that their primary and secondary projections forming the so-called interdigitating foot processes that surround the glomerular capillary loops (Garg, 2018). These foot processes are important components of GFB contributing to the high selectivity of macromolecule filtration (Haraldsson et al., 2008). In this way, as the third layer of GFB, the integrity of podocytes determines the Kf during physiological conditions. Interestingly, podocytes are also have active endocytic activity involved in the regulation of slit membrane permeability and keeping the GFB from eventual clogging (Dobrinskikh, Okamura, Kopp, Doctor, & Blaine, 2014; Inoue et al., 2019). In this section, it will be discussed the potential role of endocytosis in podocytes.

Dobrinskikh et al. (2014), using confocal imaging and total internal reflection fluorescence microscopy, showed that podocytes perform albumin uptake through caveolae-mediated endocytosis. Later, Moriyama et al. (2021), using cultured human podocytes, showed that albumin endocytosis is mediated by the neonatal Fc receptor (FcRn) followed by delivery of cargo into lysosomes or to the transcytotic or exocytotic routes. It seems that this mechanism is regulated. A previous study showed that angiotensin II (Ang II), a well-known effector component of the Renin-Angiotensin system (RAS), enhances the endocytosis and transcytosis of albumin by podocytes (Schießl et al., 2016). It has been proposed that the increase in this endocytic pathway in podocytes may be a new etiological agent for the genesis of albuminuria (Moriyama et al., 2011, 2018). In agreement with this observation, previous data also showed that Ang II infusion promotes the reduction of nephrin expression, a key protein involved in the formation of the foot process, and podocyte injury (Yang et al., 2016). Furthermore, the use of dynamin inhibitors such as sertraline (a selective serotonin reuptake inhibitor) interfered with albumin internalization through the caveolae into podocytes and reduced albuminuria in a murine model of puromycin aminonucleoside-induced nephrotic syndrome (Moriyama, Karasawa, Hasegawa, Uchida, & Nitta, 2019).

Beyond caveolae-mediated endocytosis, the podocytes also have clathrin-mediated endocytosis which is involved in the regulation of surface protein expression where implicates on glomerular structure and function (Inoue et al., 2019; Königshausen et al., 2016). It has been shown that the stimulatory effect of Ang II on glomerular permeability involves the β-arrestin-mediated nephrin endocytosis (Königshausen et al., 2016). In addition, the inhibition of type 1 angiotensin II receptor (AT1R) induces glomerular damage (Inoue et al., 2019).

Taken together, these studies allow us to suggest that podocytes can perform receptor-mediated protein endocytosis and direct the cargo through a transcytotic pathway releasing the vesicle content into urinary space. It seems this process is regulated by Ang II and other factors during physiological and pathophysiological conditions. So, some questions need to be answered: (1) what's the rate of internalized proteins directed to lysosomes or transcytotic routes? (2) what's the contribution of those transport mechanisms to those conditions? And (3) how does Ang II regulate these processes? Future studies are necessary to address those specific questions regarding the role of endocytosis and transcytosis in podocytes.

2.3 Mesangial cells and receptor-mediated endocytosis

Mesangial cells are located at the mesangium matrix among the glomerular capillaries loops and podocytes, being mainly involved in the regulation of glomerular blood flow impacting changes in glomerular filtration rate (Haraldsson et al., 2008). However, the role of mesangial in glomerular function seems to be underrated since mesangial cells can also promote clearance of macromolecules retained in the mesangium matrix working as phagocyte-like cells (Avraham, Korin, Chung, Oxburgh, & Shaw, 2021; Haraldsson et al., 2008). This strategy seems to avoid GFB clogging during physiological conditions. In this section, the potential mechanisms underlying receptor-mediated endocytosis in mesangial cells will be discussed.

In the late 70s it was reported that mesangial cells uptake and transport intravenously injected colloidal carbon particles, indicating that one of the mechanisms for clearance of materials from the mesangium was the movement of particles in the direction of the lacis area (Elema, Hoyer, & Vernier, 1976). In addition, it was shown that murine mesangial cells uptake aggregated human albumin, showing the potential role of mesangial cell endocytosis in the clearance of macromolecules (Lee & Vernier, 1980). Later, different reports described those mesangial cells cleared many substrates by phagocytosis (Cattell et al., 1982; Mancilla-Jimenez et al., 1982; Shinkai, 1982). Moreover, Bryniarski et al. (2020) showed the expression of megalin and FcRn in murine mesangial cells along with the megalin adaptor protein Dab-2. The authors also characterized the mesangial cell endocytosis of albumin showing the involvement of a receptor-mediated mechanism to be megalin-dependent.

Singhal et al. (1987), using cultured rat mesangial cells, showed that the serum macromolecules uptake involves binding to coated pits, followed by the formation of coated vesicles (endosomes), and eventually delivery of particles to lysosomes. The authors also found that an interaction between mesangial endocytosis and PGE2 production could be involved in glomerular pathophysiology. Later, Neuwirth et al. (1988) identified that this mechanism underlying PGE2 production followed by endocytosis is mediated by Fc receptors.

Rayner et al. (1990), using cultured human glomerular cells, showed that mesangial cells take up low-density lipoprotein (LDL) labeled with the fluorescent probe 1,1'-dioctadecyl-3,3,3'3'-tetramethyl-indocarbocyanine perchlorate (diI) through receptor-mediated endocytosis. Later, Wheeler et al. (1991) showed that LDL particle uptake in mesangial cells has been associated with mesangial cell lipid accumulation. In addition, this effect was stimulated by the IGF (Berfield & Abrass, 2002). These observations were correlated with a decrease in phagocytic capacity and disrupted cytoskeletons.

Koh, Nussenzveig, Okolicany, Price, and Maack (1992) showed that mesangial cells were able to internalize and promote lysosomal hydrolysis of ^{125}I-conjugated atrial natriuretic factor 1–28 (^{125}I-ANF$_{1-28}$) via clearance ANF receptors. The authors also showed that mesangial cells promote rapid receptor-mediated internalization and lysosomal hydrolysis of ^{125}I-(Sar1) Ang II. On the other hand, Singhal, Franki, Gibbons, and Hays (1992) showed that Ang II and arginine vasopressin-induced shifts in F-actin content, an important step in endocytosis, as well as the establishment of actin filament connections between mesangial cells and GBM.

Another interesting role for receptor-mediated endocytosis in mesangial cells has been identified. Using in vitro experiments, it was shown that mesangial cell endocytosis of IgG aggregates is associated with enhanced transmigration of macrophages (Singhal et al., 2000). The authors found this process is mediated by the generations of chemokines, such as RANTES and MCP-1, by mesangial cells. Teng et al. (2004) showed that mesangial cells were able to uptake light chains (LC) from light chain deposition disease (LCDD) and amyloidosis (AL) through a single receptor located at the caveolae region. However, LC endocytosis is clathrin-dependent. In addition, the intracellular trafficking in mesangial cells is different for AL-LCs and LCDD-LCs. AL-LCs are delivered to the mature lysosomal compartment where amyloid formation occurs. On the other hand, LCDD-LCs alter mesangial function and phenotype by interacting with the mesangial cell surface membranes through similar receptors as the AL-LCs.

Mesangial cells are also important in the pathophysiology of Diabetic Kidney Disease (DKD). Khera, Martin, Riley, Steadman, and Phillips (2007) showed that high glucose concentration stimulates the phagocytosis of apoptotic cells by mesangial cells. In addition, it has been shown that high glucose concentration promotes the production of TGF-β (Barro, Hsiao, Chen, Chang, & Hsieh, 2021). Furthermore, Runyan, Schnaper, and Poncelet (2005) showed that endocytosis in mesangial cells plays a critical role in TGF-β signaling, and that internalization enhances the dissociation of Smad2 from the TGF-β receptor-SARA complex, allowing Smad2 to accumulate in the nucleus and modulate target gene transcription.

Altogether, these results allow us to imagine that mesangial presents a more complex role than previously thought. Under physiological conditions, mesangial cells can perform receptor-mediated endocytosis of different factors trapped in the mesangium matrix as phagocyte-like cells. However, dysfunction of these mechanisms can be associated with the production of autacoids, pro-inflammatory chemokines, and pro-fibrotic cytokines which are observed in different kidney pathologies. In this context, some questions arise: (1) what's the molecular mechanism involved in the regulation of receptor-mediated endocytosis in mesangial cells? (2) how dysfunction of endocytosis in mesangial cells leads to glomerular injury? and (3) what is the potential impact of secretion of these inflammatory mediators in subsequent nephron segments? Future studies will help to elucidate these intriguing questions.

2.4 Parietal epithelial cells (PECs) and endocytosis

PECs form a monolayer at the internal wall of Bowman's capsule. The physiological role of PECs still needs to be elucidated. Evidence shows that PECs have a more complex role than structural ones. PECs serve as a potential precursor for podocytes, which can progressively proliferate and differentiate into podocytes to restore and maintain the number of podocytes within the glomerular tuft (Li et al., 2022). The primary cilia on the surface of PECs may play an important role in chemical and mechanical sensation, where it senses the change of flow from the glomerular filtrate and promotes the increase in intracellular calcium affecting gene expression (Ohse et al., 2009; Yoder, 2007). PECs also have contractility which can contribute to the regulation of glomerular filtration (Webber & Wong, 1973). Interestingly, there is evidence that PECs are also involved in the endocytosis of macromolecules filtered by GFB. This section will discuss the data regarding the receptor-mediated endocytosis in PECs.

Zhao et al. (2019), using primary cultured PECs, showed that albumin induces CD44 expression by PECs via the activation of the ERK signaling pathway, which is partially mediated by megalin. The authors suggested that this mechanism may be involved in glomerulosclerosis. Moreover, it seems this process involves the activation of the RAS since the treatment with losartan, a specific AT_1R antagonist, attenuated proteinuria and CD44 expression in the murine model of glomerulosclerosis.

The effect of albumin on PECs may be associated with changes in receptor-mediated endocytosis. In agreeing with this idea, Ceol et al. (2023) showed that hypertrophic PECs located at the tubular pole are potentially involved in protein endocytosis in lupus nephritis patients. The number of glomeruli presenting hypertrophic PECs positive for ClC-5, megalin, and cubilin (components associated with albumin endocytosis in renal tubular epithelial cells) was higher in lupus nephritis patients than in controls. The authors also showed that hypertrophic PECs represent a potential resource for responding to protein overload observed in other glomerulonephritis.

Regarding caveolae-mediated endocytosis in PECs, it seems to have a different perspective. Ostalska-Nowicka et al. (2007), using an immunohistochemistry approach in healthy and patients stricken with glomerular disease, showed that Cav-1 expression is high in healthy PECs. Later, this observation was confirmed by electron microscopy (Krawczyk et al., 2017). Interestingly, it was shown that Cav-1 expression is reduced in patients with focal segmental glomerulosclerosis and with lupus glomerulonephritis (Ostalska-Nowicka, Nowicki, Zachwieja, Kasper, & Witt, 2007).

Together, these results showed that PECs can promote receptor-mediated endocytosis in a megalin-dependent or caveolae-dependent manner. It seems that PECs respond to albumin overload caused by different glomerulopathies through a megalin-dependent pathway. In addition, this response is regulated by Ang II. However, there is a reduction of caveolae-dependent endocytosis. In this context, some questions should be addressed: (1) is there an interdependence between those endocytic pathways? (2) Is Ang II mediating the decrease in caveolae-dependent endocytosis? (3) what's the potential consequence of activated PECs by protein overload in subsequent nephron portions? (4) Does the hypertrophy of PECs involve a generation of secretory phenotype contributing to tubular injury? Future studies are necessary to elucidate those questions.

3. Proximal tubule epithelial cells and endocytosis

The proximal tubule (PT) is the nephron segment responsible for the reabsorption of almost 70% of water and solutes from tubular fluid in a coupled manner (Feraille, Sassi, Olivier, Arnoux, & Martin, 2022; McDonough & Layton, 2023). The PT epithelial cells (PTECs) are polarized epithelial cells that present a large number of mitochondria in the basolateral regions associated with high production of ATP and active transport. The energy expended by primary active transporters generates the electrochemical gradient used for Na^+-dependent secondary active transporters located at the luminal membrane. This is the main mechanism involved in the reabsorption of essential electrolytes and organic solutes such as amino acids, glucose, and several metabolites. Tubular reabsorption of water occurs in an isosmotic manner due to the high expression of water channels known as type 1 aquaporin. In addition to the transport tasks, the PTECs are important sensors, mediators, and effectors of many functions by secreting hormones such as active D vitamin or erythropoietin, and others. Interestingly, the PTECs are cells that possess a well-developed endocytic apparatus that are involved in the reabsorption of proteins filtered by the glomerulus (Polesel & Hall, 2019). In this section, we will discuss the mechanisms involved in this process as well as their regulatory mechanisms under physiological and pathophysiological conditions.

3.1 Proximal tubule epithelial cells and receptor-mediated endocytosis

In the middle 80s Park and Maack (1984), using isolated rabbit PT perfused with radioactively labeled albumin, showed that PTECs transport albumin in a saturable process suggesting a receptor-dependent transport mechanism. Beyond that, the authors showed that the mechanism presented dual kinetics: low- and high-capacity systems. Later, Schwegler, Heppelmann, Mildenberger, and Silbernagl (1991) showed that opossum kidney (OK) cells, a PTEC line, promote the internalization of fluorescein isothiocyanate-conjugated albumin due to receptor-mediated endocytosis. Using an ultrastructural analysis of PTEC incubated with labeled albumin, Gómez-Pascual, Londoño, Ghitescu, Desjardins, and Bendayan (1995) showed that in the PTECs, labeling was present over microvilli as well as over endosomal and lysosomal compartments, with labeling intensities varying from one compartment to the other. In addition, Zhai et al. (2000) showed that albumin endocytosis is mediated by megalin and cubilin. At the same time, other PTECs linage were shown to promote protein endocytosis such as LLC-PK1 cells (Takakura et al., 1995). Furthermore, Nielsen et al. (1998) showed that LLC-PK1 cells also present high expression of megalin.

Evidence also described a retrieval pathway associated with intact albumin transport. Russo et al. (2007), using fluorescent labeled albumin and a two-photon microscopy technique, showed albumin is partially reabsorbed by transcytosis in PTECs. Tenten et al. (2013) showed that this process is involved in the rescue of albumin and IgGs from the degradation pathway in PTECs. The authors also showed that this process involves the activity of the FcRn. Furthermore, a defect in this mechanism was associated with hypoalbuminemia.

Beyond albumin itself, PTECs can perform the endocytosis of different ligands. Gburek et al. (2002) also showed that these receptors are involved in hemoglobin endocytosis. In addition, using OK cells, Ly, Tesch, Nikolic-Paterson, and Poronnik (2019) showed that PTECs preferentially endocytose covalently modified albumin compared to native albumin. This apparent selectivity of the scavenger receptor complex suggests a specific role for this pathway in the removal of modified albumin from the bloodstream.

Other receptors are involved in protein endocytosis by PTECs. Iwao et al. (2008) showed that CD36, a transmembrane protein of a class B scavenger receptor, which is expressed in the human kidney 2 cell line (HK-2 cells), can internalize albumin and advanced oxidation protein products. The authors proposed a potential role of CD36-mediated endocytosis in the development of renal tubular injury. Interestingly, it has been shown that kidney injury molecule-1, a scavenger receptor usually expressed in PTECs during pathological conditions and a specific tubular injury marker, promotes internalization of albumin PTECs, at least partially, via a clathrin-dependent mechanism (Zhao, Jiang, Olufade, Liu, & Emmett, 2016).

The mechanism underlying the receptor-mediated endocytosis of albumin has been studied. The inhibition of isoform 3 of Na+/H+ exchanger (NHE3) activity decreases albumin internalization impairing receptor-mediated albumin endocytosis in OK cells (Gekle et al., 1999). Later, it was shown that this effect was related to the modulation of the endocytic vesical fusion (Gekle, Freudinger, & Mildenberger, 2001). Beyond NHE3, it was also shown that albumin endocytosis by PTECs involves the interaction between the C-terminal tail of ClC-5 and cofilin, an actin-associated protein (Hryciw et al., 2003). Furthermore, the interaction between ClC-5 and albumin endocytic mechanisms is dependent on the NHERF1 and NHERF2 (Hryciw et al., 2006).

Receptor-mediated albumin endocytosis is described to be regulated by multiple factors and intracellular signaling proteins. Hryciw et al. (2005) showed that PKC-α is involved in constitutive albumin uptake in OK cells by mediating the assembly of actin microfilaments at the apical membrane. In addition, Gekle et al. (1997) showed that albumin uptake in OK cells is also dependent on PKA activity. Furthermore, Grieco et al. (2018) showed that Vps34/PI3-kinase type III is a key in vivo component of molecular machinery governing apical vesicular trafficking and, thus absorptive function in PTECs. Defects in the Vps34 signaling were associated with the development of Fanconi-*like* syndrome.

Albumin endocytosis by PTECs is also regulated by the Akt/PKB activity (Caruso-Neves, Kwon, & Guggino, 2005; Coffey, Costacou, Orchard, & Erkan, 2015; Koral et al., 2014; Peruchetti, Silva-Aguiar, Siqueira, Dias, & Caruso-Neves, 2018; Silva-Aguiar et al., 2022; Teixeira et al., 2019). Using LLC-PK1 cells, Caruso-Neves, Pinheiro, Cai, Souza-Menezes, and Guggino (2006) showed a pool of Akt bound to the intracellular region of megalin in the basal condition. In this work, the authors showed, under physiological conditions, that albumin binding to megalin results in Akt phosphorylation and activation, and consequently, the detachment from megalin finally translocates to cytosol and develops their functions. One of them was to positively regulate their albumin endocytosis by stimulating megalin recycling to the luminal membrane (Silva-Aguiar et al., 2022). Beyond the albumin effects, it has been shown that insulin increased albumin endocytosis in PTECs in an Akt-dependent mechanism (Coffey et al., 2015). On the other hand, inhibition of Akt activity during pathological conditions, such as subclinical acute kidney injury (subAKI) and DKD, we observed a reduction of megalin-dependent albumin endocytosis in PTECs (Peruchetti et al., 2018; Teixeira et al., 2019).

Beyond the direct effect of Akt on megalin-dependent albumin endocytosis, there is evidence involving the modulation of the mTORC1 pathway. In mammals, mTOR is a serine/threonine protein kinase that forms distinct protein complexes such as mTORC1 and mTORC2 (Fantus, Rogers, Grahammer, Huber, & Thomson, 2016). Akt can directly modulate the mTOR pathway in many cell types, including PTECs (Fantus et al., 2016). It has been described that mTOR, a serine/threonine kinase, is important to maintain the basal megalin-dependent albumin endocytosis by PTECs (Oroszlán et al., 2010; Peres et al., 2023). Evidence points out that disruption of mTOR signaling leads to the development of tubular proteinuria (Coombes, Mreich, Liddle, & Rangan, 2005; Peres et al., 2023). On the other hand, disruption of megalin-mediated albumin endocytosis enhances mTORC1 activity and the development of tubule-interstitial injury (Peruchetti, Cheng, Caruso-Neves, & Guggino, 2014).

Other intracellular proteins are involved in the regulation of albumin endocytosis by PTECs. Terryn et al. (2016) showed that hepatocyte nuclear factor 1α plays a key role in the constitutive expression of megalin and cubilin, hence regulating PT endocytosis. In addition, Zeng et al. (2017) showed that stromal interaction molecule-1/calcium release-activated calcium modulator 1 (ORAI1) colocalizes with clathrin, but not with caveolin, at the apical membrane of PTECs, which determines clathrin-mediated endocytosis. The authors proposed that inhibition of ORAI expression induced by hyperglycemia is one of the possible mechanisms involved in the development of albuminuria observed in DKD.

Different works showed that PTECs receptor-mediated albumin endocytosis is modulated by several stimuli, including pro-fibrotic cytokines. Gekle et al. (2003) showed that transforming growth factor-β1 (TGF-β1) reduces megalin- and cubilin-mediated endocytosis of albumin in OK cells. The authors found that TGF-β1 leads to Smad2- and Smad3-dependent expression of negative regulators of receptor-mediated endocytosis. In addition, the dysfunction of PTEC albumin endocytosis induced by protein overload increases TGF-β secretion in the renal cortex (Landgraf et al., 2014; Peruchetti et al., 2020). Another possibility is the fluid flow through the tubular lumen. Raghavan, Rbaibi, Pastor-Soler, Carattino, and Weisz (2014) showed that fluid shear stress modulates apical receptor-mediated endocytosis by microvillar bending that contributes to the efficient retrieval of filtered proteins in the proximal tubules. Later, Gualdani et al. (2020) showed that the ion channel Transient Receptor Potential Cation Channel Subfamily V Member 4 (TRPV4) modulated the endocytosis of albumin and low-molecular-weight proteins in the proximal tubule. The authors also showed the importance of TRPV4 in sensing pressure in the PT in response to variations in the amount of ultrafiltrate and unveiled a mechanism that controls protein reabsorption.

Interestingly, the receptor-mediated albumin endocytosis by PTECs is also mediated by vasoactive peptides. Caruso-Neves et al. (2005), using LLC-PK1 cells, showed that Ang II stimulates albumin endocytosis due to the activation of the Ang II type 2 receptor. On the other hand, Peruchetti et al. (2021) showed that Ang II through the activation of AT_1R promotes the inhibition of megalin-mediated endocytosis contributing to the development of tubular proteinuria in subAKI. More recently, Alves et al. (2021) showed that bradykinin, a well-known vasodilator and natriuretic agent, through the activation of bradykinin B2 receptor decreases megalin-mediated albumin endocytosis in LLC-PK1 cells.

Albumin endocytosis by PTECs is impaired during different patho-physiological conditions. Choi, Kim, Ahn, and Park (1999) showed that OK cells exposed to cadmium chloride presented reduced albumin endocytosis. In addition, Fujishiro et al. (2020) showed that Cd affects the tubular reabsorption of low-molecular-weight proteins even at nonlethal concentrations. Furthermore, it has been shown that cadmium impairs albumin reabsorption by down-regulating megalin and ClC5 channels in LLC-PK1 cells (Gena, Calamita, & Guggino, 2010). Moreover, Takano et al. (2002) showed that cisplatin also decreases the receptor-mediated endocytosis of protein following the inhibition of vacuolar H+-ATPase in OK cells, and the inhibition of receptor-mediated endocytosis would be the mechanisms underlying the proteinuria induced by cisplatin treatment. Genetic diseases are also impacting albumin endocytosis by PTECs. For example, Gorvin et al. (2013) showed that receptor-mediated endocytosis and endosomal acidification are impaired in PTECs of Dent disease patients, a syndrome generated by ClC-5 mutations. The development of tubular proteinuria caused by infections-induced kidney disease has been observed. In this context, Silva-Aguiar et al. (2022) showed that the inhibition of albumin endocytosis by isolated S protein, a key protein of

SARS-CoV-2 infection, inhibits megalin expression without changes in its trafficking and stability. These results reveal a possible mechanism to explain the albuminuria observed in patients stricken with COVID-19 (Werion et al., 2020).

Some reports show an association between the dysfunction of albumin endocytosis and the development of tubular injury as well as renal inflammation and fibrosis. Caruso-Neves et al. (2006) showed that albumin overload results in increased cell apoptosis in LLC-PK1 cells. In agreement, it was shown that protein overload increased PTECs apoptosis (Thomas et al., 1999). Regarding the association between albumin endocytosis and renal pro-inflammatory phenotype, it has been shown that high albumin endocytosis is associated with the generation of ROS and IL-8 secretion (Tang et al., 2003). In addition, Drumm, Gassner, Silbernagl, and Gekle (2001) showed that albumin overload promotes NF-kappa B activation in OK and LLC-PK1 cells. Furthermore, Pearson, Colville-Nash, Kwan, and Dockrell (2008) showed that albumin overload-induced IL-6 secretion by primary human PTECs.

Beyond the pro-inflammatory phenotype, it was also shown that excessive protein uptake by PTECs is involved in disruption in collagen homeostasis resulting in probable progression of interstitial fibrosis (Wohlfarth, Drumm, Mildenberger, Freudinger, & Gekle, 2003). In addition, Diwakar, Pearson, Colville-Nash, Brunskill, and Dockrell (2007) showed that albumin overload induces secretion of TGF-β1 in human kidney cell clone 8 (HKC-8) and OK cells. Furthermore, Chen et al. (2017) showed that albumin endocytosis leads to an increase in matrix metalloproteinase 9, a protein dysregulated in chronic kidney diseases such as DKD, which is associated with kidney fibrosis.

Altogether, these data allow us to propose that PTECs are important to the reabsorption of albumin filtered into the glomerulus. The mechanism involved in this process is dependent on receptor-mediated endocytosis where megalin, along with other scavenger receptors, is responsible for interacting with albumin and other ligands at the tubular lumen. The mechanism is tightly regulated by different signaling molecules and associated intracellular signaling. The dysregulation of receptor-mediated albumin endocytosis is observed in many kidney pathologies where is observed proteinuria and albuminuria as well as tubule-interstitial injury. However, important questions arise regarding this process: (1) what's the role of albumin endocytosis for PTEC biology during physiological conditions; (2) what's the molecular mechanism involved in this process? (3) what's the role of impaired receptor-mediated albumin endocytosis during kidney pathologies? Future experiments will help to clarify these issues.

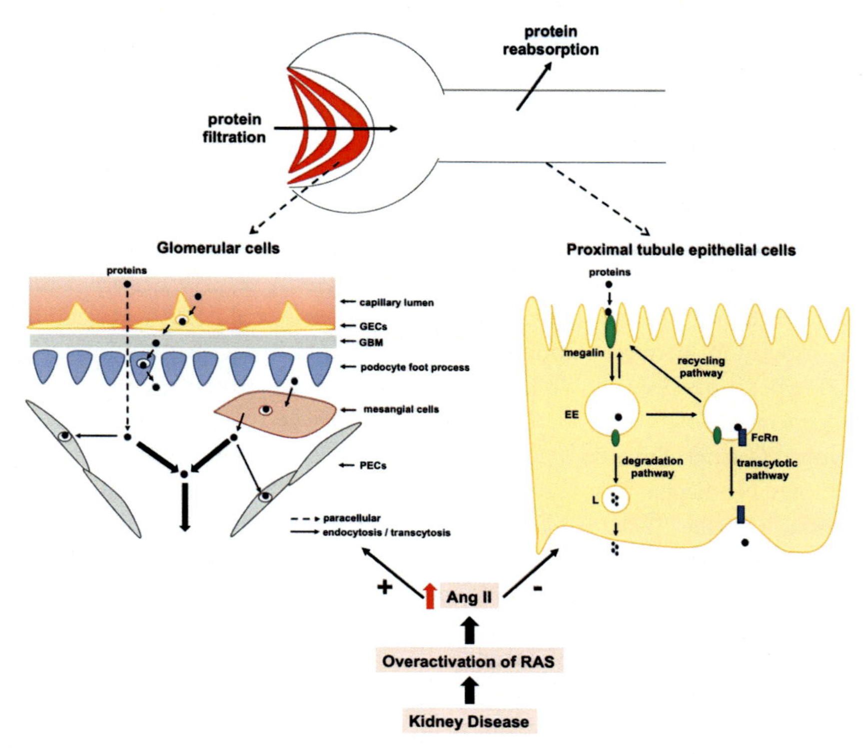

Fig. 1 Receptor-mediated endocytosis in kidney cells during physiological and pathological conditions (details described in text). GECs, glomerular endothelial cells; GBM, glomerular basal membrane; PECs, parietal epithelial cells; EE, early endosomes; L, lysosomes; FcRn, neonatal Fc receptor; Ang II, angiotensin II; RAS, Renin-Angiotensin System.

4. Conclusions and future perspectives

Altogether, these findings show that renal handling of proteins is more complex than been thought. All results allow us to propose a model for the renal handling of proteins in physiological and pathological conditions (Fig. 1). Previously, it has been described that low-molecular-weight proteins and albumin are filtered into the glomerulus through paracellular transport across the GFB. New evidence is pointing out that filtration of these macromolecules may also occur through receptor-mediated endocytosis followed by transcytotic pathways in glomerular cells such as GECs and podocytes. In addition, rather than controlling the glomerular capillary flow, the mesangial cells can perform endocytosis of albumin to avoid the potential clog of GFB contributing to clearing the GBM from aggregate deposits. Once reaching the urinary space, low-molecular-weight proteins and albumin can be internalized through receptor-mediated endocytosis by PECs. When it appears in the tubular lumen, the filtered proteins are promptly reabsorbed by PTECs through receptor-mediated endocytosis. On these cells, the internalized albumin can go through the lysosomal degradation pathway or through the transcytosis pathway, also known as the retrieval pathway. All these processes in the nephron compartments are regulated. Evidence points out the role of RAS components, mainly Ang II, in the regulation of receptor-mediated albumin endocytosis in both glomerular and tubular cells. Exacerbation of Ang II levels is associated with the development of proteinuria and kidney injury. The potential mechanisms involved in this process could be Ang II acting directly on kidney cells leading to: (1) an increase in the endocytosis and transcytosis of albumin through the different glomerular cell types; and (2) the decrease in endocytosis by PTECs. The importance of these mechanisms is reinforced due to the protective effect of angiotensin receptor blockers as anti-proteinuric agents in different kidney diseases. The future investigation of the endocytic routes as well as the molecular mechanisms underlying these processes will be an attractive option to elucidate the development of tubular injuries and proteinuria. Furthermore, it will be important to identify new markers of injury and new potential therapeutic targets.

Acknowledgments

This work has been funded by: (1) Fundação Amparo à Pesquisa do Estado de Minas Gerais (FAPEMIG): PROBIC-FAPEMIG scholarship for MRC. (2) Pró-Reitoria de Pesquisa (PRPq) of the Federal University of Minas Gerais (UFMG) for DBP. (3) Conselho Nacional de Desenvolvimento Científico e Tecnológico (CNPq): INCT-NANOBiofar-CNPq (#406792/2022-4) for DBP; PIBIT-CNPq scholarship (#167458/2023-0) for LBFO.

Conflict of interest

The authors declare that there is no conflict of interest.

References

Abbate, M., Zoja, C., & Remuzzi, G. (2006). How does proteinuria cause progressive renal damage? *Journal of the American Society of Nephrology: JASN, 17*(11), 2974–2984. https://doi.org/10.1681/ASN.2006040377.

Alves, S. A. S., Florentino, L. S., Teixeira, D. E., Silva-Aguiar, R. P., Peruchetti, D. B., Oliveira, A. C., ... Caruso-Neves, C. (2021). Surface megalin expression is a target to the inhibitory effect of bradykinin on the renal albumin endocytosis. *Peptides, 146*, 170646. https://doi.org/10.1016/j.peptides.2021.170646.

Avraham, S., Korin, B., Chung, J.-J., Oxburgh, L., & Shaw, A. S. (2021). The Mesangial cell—The glomerular stromal cell. *Nature Reviews. Nephrology, 17*(12), 855–864. https://doi.org/10.1038/s41581-021-00474-8.

Barro, L., Hsiao, J.-T., Chen, C.-Y., Chang, Y.-L., & Hsieh, M.-F. (2021). Cytoprotective effect of liposomal puerarin on high glucose-induced injury in rat mesangial cells. *Antioxidants (Basel, Switzerland), 10*(8), 1177. https://doi.org/10.3390/antiox10081177.

Berfield, A. K., & Abrass, C. K. (2002). IGF-1 induces foam cell formation in rat glomerular mesangial cells. *Journal of Histochemistry & Cytochemistry, 50*(3), 395–403. https://doi.org/10.1177/002215540205000310.

Bryniarski, M. A., Yee, B. M., Chaves, L. D., Stahura, C. M., Yacoub, R., & Morris, M. E. (2020). Megalin-mediated albumin endocytosis in cultured murine mesangial cells. *Biochemical and Biophysical Research Communications, 529*(3), 740–746. https://doi.org/10.1016/j.bbrc.2020.05.166.

Caruso-Neves, C., Kwon, S.-H., & Guggino, W. B. (2005). Albumin endocytosis in proximal tubule cells is modulated by angiotensin II through an AT2 receptor-mediated protein kinase B activation. *Proceedings of the National Academy of Sciences of the United States of America, 102*(48), 17513–17518. https://doi.org/10.1073/pnas.0507255102.

Caruso-Neves, C., Pinheiro, A. A. S., Cai, H., Souza-Menezes, J., & Guggino, W. B. (2006). PKB and megalin determine the survival or death of renal proximal tubule cells. *Proceedings of the National Academy of Sciences of the United States of America, 103*(49), 18810–18815. https://doi.org/10.1073/pnas.0605029103.

Cattell, V., Gaskin de Urdaneta, A., Arlidge, S., Collar, J. E., Roberts, A., & Smith, J. (1982). Uptake and clearance of ferritin by the glomerular mesangium. I. Phagocytosis by mesangial cells and blood monocytes. *Laboratory Investigation; A Journal of Technical Methods and Pathology, 47*(3), 296–303.

Ceol, M., Gianesello, L., Trimarchi, H., Migliorini, A., Priante, G., Radu, C. M., ... Del Prete, D. (2023). Human parietal epithelial cells (PECs) and proteinuria in lupus nephritis: A role for ClC-5, megalin, and cubilin? *Journal of Nephrology*. https://doi.org/10.1007/s40620-023-01725-6.

Chen, X., Cobbs, A., George, J., Chima, A., Tuyishime, F., & Zhao, X. (2017). Endocytosis of albumin induces matrix metalloproteinase-9 by activating the ERK signaling pathway in renal tubule epithelial cells. *International Journal of Molecular Sciences, 18*(8), 1758. https://doi.org/10.3390/ijms18081758.

Choi, J. S., Kim, K. R., Ahn, D. W., & Park, Y. S. (1999). Cadmium inhibits albumin endocytosis in opossum kidney epithelial cells. *Toxicology and Applied Pharmacology, 161*(2), 146–152. https://doi.org/10.1006/taap.1999.8797.

Coffey, S., Costacou, T., Orchard, T., & Erkan, E. (2015). Akt links insulin signaling to albumin endocytosis in proximal tubule epithelial cells. *PLoS One, 10*(10), e0140417. https://doi.org/10.1371/journal.pone.0140417.

Coombes, J. D., Mreich, E., Liddle, C., & Rangan, G. K. (2005). Rapamycin worsens renal function and intratubular cast formation in protein overload nephropathy. *Kidney International, 68*(6), 2599–2607. https://doi.org/10.1111/j.1523-1755.2005.00732.x.

Cooper, S. T., & McNeil, P. L. (2015). Membrane repair: Mechanisms and pathophysiology. *Physiological Reviews, 95*(4), 1205–1240. https://doi.org/10.1152/physrev.00037.2014.

Dickson, L. E., Wagner, M. C., Sandoval, R. M., & Molitoris, B. A. (2014). The proximal tubule and albuminuria: Really!. *Journal of the American Society of Nephrology, 25*(3), 443–453. https://doi.org/10.1681/ASN.2013090950.

Diwakar, R., Pearson, A. L., Colville-Nash, P., Brunskill, N. J., & Dockrell, M. E. C. (2007). The role played by endocytosis in albumin-induced secretion of TGF-beta1 by proximal tubular epithelial cells. *American Journal of Physiology. Renal Physiology, 292*(5), F1464–F1470. https://doi.org/10.1152/ajprenal.00069.2006.

Dobrinskikh, E., Okamura, K., Kopp, J. B., Doctor, R. B., & Blaine, J. (2014). Human podocytes perform polarized, caveolae-dependent albumin endocytosis. *American Journal of Physiology. Renal Physiology, 306*(9), F941–F951. https://doi.org/10.1152/ajprenal.00532.2013.

Drumm, K., Gassner, B., Silbernagl, S., & Gekle, M. (2001). Inhibition of Na superset+/H superset+ exchange decreases albumin-induced NF-kappaB activation in renal proximal tubular cell lines (OK and LLC-PK1 cells). *European Journal of Medical Research, 6*(10), 422–432.

Elema, J. D., Hoyer, J. R., & Vernier, R. L. (1976). The glomerular mesangium: Uptake and transport of intravenously injected colloidal carbon in rats. *Kidney International, 9*(5), 395–406. https://doi.org/10.1038/ki.1976.49.

Fantus, D., Rogers, N. M., Grahammer, F., Huber, T. B., & Thomson, A. W. (2016). Roles of mTOR complexes in the kidney: Implications for renal disease and transplantation. *Nature Reviews. Nephrology, 12*(10), 587–609. https://doi.org/10.1038/nrneph.2016.108.

Feraille, E., Sassi, A., Olivier, V., Arnoux, G., & Martin, P.-Y. (2022). Renal water transport in health and disease. *Pflugers Archiv: European Journal of Physiology, 474*(8), 841–852. https://doi.org/10.1007/s00424-022-02712-9.

Fujishiro, H., Yamamoto, H., Otera, N., Oka, N., Jinno, M., & Himeno, S. (2020). In vitro evaluation of the effects of cadmium on endocytic uptakes of proteins into cultured proximal tubule epithelial cells. *Toxics, 8*(2), 24. https://doi.org/10.3390/toxics8020024.

Garg, P. (2018). A review of podocyte biology. *American Journal of Nephrology, 47*(Suppl. 1), 3–13. https://doi.org/10.1159/000481633.

Gburek, J., Verroust, P. J., Willnow, T. E., Fyfe, J. C., Nowacki, W., Jacobsen, C., ... Christensen, E. I. (2002). Megalin and cubilin are endocytic receptors involved in renal clearance of hemoglobin. *Journal of the American Society of Nephrology: JASN, 13*(2), 423–430. https://doi.org/10.1681/ASN.V132423.

Gekle, M., Drumm, K., Mildenberger, S., Freudinger, R., Gassner, B., & Silbernagl, S. (1999). Inhibition of Na+-H+ exchange impairs receptor-mediated albumin endocytosis in renal proximal tubule-derived epithelial cells from opossum. *The Journal of Physiology, 520*(Pt 3), 709–721. https://doi.org/10.1111/j.1469-7793.1999.00709.x.

Gekle, M., Freudinger, R., & Mildenberger, S. (2001). Inhibition of Na+-H+ exchanger-3 interferes with apical receptor-mediated endocytosis via vesicle fusion. *The Journal of Physiology, 531*(Pt 3), 619–629. https://doi.org/10.1111/j.1469-7793.2001.0619h.x.

Gekle, M., Mildenberger, S., Freudinger, R., Schwerdt, G., & Silbernagl, S. (1997). Albumin endocytosis in OK cells: Dependence on actin and microtubules and regulation by protein kinases. *The American Journal of Physiology, 272*(5 Pt 2), F668–F677. https://doi.org/10.1152/ajprenal.1997.272.5.F668.

Gekle, M. (2005). Renal tubule albumin transport. *Annual Review of Physiology, 67*(1), 573–594. https://doi.org/10.1146/annurev.physiol.67.031103.154845.

Gekle, M., Knaus, P., Nielsen, R., Mildenberger, S., Freudinger, R., Wohlfarth, V., ... Christensen, E. I. (2003). Transforming growth factor-beta1 reduces megalin- and cubilin-mediated endocytosis of albumin in proximal-tubule-derived opossum kidney cells. *The Journal of Physiology, 552*(Pt 2), 471–481. https://doi.org/10.1113/jphysiol.2003.048074.

Gena, P., Calamita, G., & Guggino, W. B. (2010). Cadmium impairs albumin reabsorption by down-regulating megalin and ClC5 channels in renal proximal tubule cells. *Environmental Health Perspectives, 118*(11), 1551–1556. https://doi.org/10.1289/ehp.0901874.

Gómez-Pascual, A., Londoño, I., Ghitescu, L., Desjardins, M., & Bendayan, M. (1995). Immunocytochemical investigation of the in vivo endocytosis by renal tubular epithelial cells. *Microscopy Research and Technique, 31*(2), 118–127. https://doi.org/10.1002/jemt.1070310204.

Gorvin, C. M., Wilmer, M. J., Piret, S. E., Harding, B., van den Heuvel, L. P., Wrong, O., ... Thakker, R. V. (2013). Receptor-mediated endocytosis and endosomal acidification is impaired in proximal tubule epithelial cells of Dent disease patients. *Proceedings of the National Academy of Sciences of the United States of America, 110*(17), 7014–7019. https://doi.org/10.1073/pnas.1302063110.

Grieco, G., Janssens, V., Gaide Chevronnay, H. P., N'Kuli, F., Van Der Smissen, P., Wang, T., ... Courtoy, P. J. (2018). Vps34/PI3KC3 deletion in kidney proximal tubules impairs apical trafficking and blocks autophagic flux, causing a Fanconi-like syndrome and renal insufficiency. *Scientific Reports, 8*(1), 14133. https://doi.org/10.1038/s41598-018-32389-z.

Gualdani, R., Seghers, F., Yerna, X., Schakman, O., Tajeddine, N., Achouri, Y., ... Gailly, P. (2020). Mechanical activation of TRPV4 channels controls albumin reabsorption by proximal tubule cells. eabc6967 *Science Signaling, 13*(653), https://doi.org/10.1126/scisignal.abc6967.

Haraldsson, B., Nyström, J., & Deen, W. M. (2008). Properties of the glomerular barrier and mechanisms of proteinuria. *Physiological Reviews, 88*(2), 451–487. https://doi.org/10.1152/physrev.00055.2006.

Hryciw, D. H., Ekberg, J., Ferguson, C., Lee, A., Wang, D., Parton, R. G., ... Poronnik, P. (2006). Regulation of albumin endocytosis by PSD95/Dlg/ZO-1 (PDZ) scaffolds. Interaction of Na+-H+ exchange regulatory factor-2 with ClC-5. *The Journal of Biological Chemistry, 281*(23), 16068–16077. https://doi.org/10.1074/jbc.M512559200.

Hryciw, D. H., Pollock, C. A., & Poronnik, P. (2005). PKC-alpha-mediated remodeling of the actin cytoskeleton is involved in constitutive albumin uptake by proximal tubule cells. *American Journal of Physiology. Renal Physiology, 288*(6), F1227–F1235. https://doi.org/10.1152/ajprenal.00428.2003.

Hryciw, D. H., Wang, Y., Devuyst, O., Pollock, C. A., Poronnik, P., & Guggino, W. B. (2003). Cofilin interacts with ClC-5 and regulates albumin uptake in proximal tubule cell lines. *The Journal of Biological Chemistry, 278*(41), 40169–40176. https://doi.org/10.1074/jbc.M307890200.

Inoue, K., Tian, X., Velazquez, H., Soda, K., Wang, Z., Pedigo, C. E., ... Ishibe, S. (2019). Inhibition of endocytosis of clathrin-mediated angiotensin II receptor type 1 in podocytes augments glomerular injury. *Journal of the American Society of Nephrology: JASN, 30*(12), 2307–2320. https://doi.org/10.1681/ASN.2019010053.

Iwao, Y., Nakajou, K., Nagai, R., Kitamura, K., Anraku, M., Maruyama, T., & Otagiri, M. (2008). CD36 is one of important receptors promoting renal tubular injury by advanced oxidation protein products. *American Journal of Physiology. Renal Physiology, 295*(6), F1871–F1880. https://doi.org/10.1152/ajprenal.00013.2008.

Khera, T. K., Martin, J., Riley, S. G., Steadman, R., & Phillips, A. O. (2007). Glucose modulates handling of apoptotic cells by mesangial cells: Involvement of TGF-β1. *Laboratory Investigation, 87*(7), 690–701. https://doi.org/10.1038/labinvest.3700555.

Koh, G. Y., Nussenzveig, D. R., Okolicany, J., Price, D. A., & Maack, T. (1992). Dynamics of atrial natriuretic factor-guanylate cyclase receptors and receptor-ligand complexes in cultured glomerular mesangial and renomedullary interstitial cells. *Journal of Biological Chemistry, 267*(17), 11987–11994. https://doi.org/10.1016/S0021-9258(19)49795-0.

Königshausen, E., Zierhut, U. M., Ruetze, M., Potthoff, S. A., Stegbauer, J., Woznowski, M., ... Sellin, L. (2016). Angiotensin II increases glomerular permeability by β-arrestin mediated nephrin endocytosis. *Scientific Reports, 6*, 39513. https://doi.org/10.1038/srep39513.

Koral, K., Li, H., Ganesh, N., Birnbaum, M. J., Hallows, K. R., & Erkan, E. (2014). Akt recruits Dab2 to albumin endocytosis in the proximal tubule. *American Journal of Physiology. Renal Physiology, 307*(12), F1380–F1389. https://doi.org/10.1152/ajprenal.00454.2014.

Krawczyk, K. M., Hansson, J., Nilsson, H., Krawczyk, K. K., Swärd, K., & Johansson, M. E. (2017). Injury induced expression of caveolar proteins in human kidney tubules—Role of megakaryoblastic leukemia 1. *BMC Nephrology, 18*(1), 320. https://doi.org/10.1186/s12882-017-0738-8.

Landgraf, S. S., Silva, L. S., Peruchetti, D. B., Sirtoli, G. M., Moraes-Santos, F., Portella, V. G., ... Caruso-Neves, C. (2014). 5-Lypoxygenase products are involved in renal tubulointerstitial injury induced by albumin overload in proximal tubules in mice. *PLoS One, 9*(10), e107549. https://doi.org/10.1371/journal.pone.0107549.

Lee, S., & Vernier, R. L. (1980). Immunoelectron microscopy of the glomerular mesangial uptake and transport of aggregated human albumin in the mouse. *Laboratory Investigation; a Journal of Technical Methods and Pathology, 42*(1), 44–58.

Li, Z., Guo, X., Quan, X., Yang, C., Liu, Z., Su, H., ... Liu, H. (2022). The role of parietal epithelial cells in the pathogenesis of podocytopathy. *Frontiers in Physiology, 13*, 832772. https://doi.org/10.3389/fphys.2022.832772.

Ly, N. D., Tesch, G. H., Nikolic-Paterson, D. J., & Poronnik, P. (2019). Proximal tubular epithelial cells preferentially endocytose covalently-modified albumin compared to native albumin. *Nephrology (Carlton, Vic.), 24*(1), 121–126. https://doi.org/10.1111/nep.13211.

Mancilla-Jimenez, R., Bellon, B., Kuhn, J., Belair, M. F., Rouchon, M., Druet, P., & Bariety, J. (1982). Phagocytosis of heat-aggregated immunoglobulins by mesangial cells: An immunoperoxidase and acid phosphatase study. *Laboratory Investigation; a Journal of Technical Methods and Pathology, 46*(3), 243–253.

McDonough, A. A., & Layton, A. T. (2023). Sex differences in renal electrolyte transport. *Current Opinion in Nephrology and Hypertension, 32*(5), 467–475. https://doi.org/10.1097/MNH.0000000000000909.

McMahon, H. T., & Boucrot, E. (2011). Molecular mechanism and physiological functions of clathrin-mediated endocytosis. *Nature Reviews. Molecular Cell Biology, 12*(8), 517–533. https://doi.org/10.1038/nrm3151.

Moeller, M. J., & Tenten, V. (2013). Renal albumin filtration: Alternative models to the standard physical barriers. *Nature Reviews Nephrology, 9*(5), 266–277. https://doi.org/10.1038/nrneph.2013.58.

Moriyama, T., Hasegawa, F., Miyabe, Y., Akiyama, K., Karasawa, K., Uchida, K., & Nitta, K. (2021). Intracellular trafficking pathway of albumin in glomerular epithelial cells. *Biochemical and Biophysical Research Communications, 574*, 97–103. https://doi.org/10.1016/j.bbrc.2021.08.043.

Moriyama, T., Karasawa, K., Hasegawa, F., Uchida, K., & Nitta, K. (2019). Sertraline reduces albuminuria by interfering with caveolae-mediated endocytosis through glomerular endothelial and epithelial cells. *American Journal of Nephrology, 50*(6), 444–453. https://doi.org/10.1159/000503917.

Moriyama, T., Karasawa, K., & Nitta, K. (2018). The role of caveolae on albumin passage through glomerular endothelial and epithelial cells: The new etiology of urinary albumin excretion. *Contributions to Nephrology, 195*, 1–11. https://doi.org/10.1159/000486929.

Moriyama, T., Sasaki, K., Karasawa, K., Uchida, K., & Nitta, K. (2017). Intracellular transcytosis of albumin in glomerular endothelial cells after endocytosis through caveolae. *Journal of Cellular Physiology, 232*(12), 3565–3573. https://doi.org/10.1002/jcp.25817.

Moriyama, T., Takei, T., Itabashi, M., Uchida, K., Tsuchiya, K., & Nitta, K. (2015). Caveolae may enable albumin to enter human renal glomerular endothelial cells. *Journal of Cellular Biochemistry, 116*(6), 1060–1069. https://doi.org/10.1002/jcb.25061.

Moriyama, T., Tsuruta, Y., Shimizu, A., Itabashi, M., Takei, T., Horita, S., ... Nitta, K. (2011). The significance of caveolae in the glomeruli in glomerular disease. *Journal of Clinical Pathology, 64*(6), 504–509. https://doi.org/10.1136/jcp.2010.087023.

Neuwirth, R., Singhal, P., Diamond, B., Hays, R. M., Lobmeyer, L., Clay, K., & Schlondorff, D. (1988). Evidence for immunoglobulin Fc receptor-mediated prostaglandin2 and platelet-activating factor formation by cultured rat mesangial cells. *The Journal of Clinical Investigation, 82*(3), 936–944. https://doi.org/10.1172/JCI113701.

Nielsen, R., Birn, H., Moestrup, S. K., Nielsen, M., Verroust, P., & Christensen, E. I. (1998). Characterization of a kidney proximal tubule cell line, LLC-PK1, expressing endocytotic active megalin. *Journal of the American Society of Nephrology: JASN, 9*(10), 1767–1776. https://doi.org/10.1681/ASN.V9101767.

Ohse, T., Pippin, J. W., Chang, A. M., Krofft, R. D., Miner, J. H., Vaughan, M. R., & Shankland, S. J. (2009). The enigmatic parietal epithelial cell is finally getting noticed: A review. *Kidney International, 76*(12), 1225–1238. https://doi.org/10.1038/ki.2009.386.

Oroszlán, M., Bieri, M., Ligeti, N., Farkas, A., Meier, B., Marti, H.-P., & Mohacsi, P. (2010). Sirolimus and everolimus reduce albumin endocytosis in proximal tubule cells via an angiotensin II-dependent pathway. *Transplant Immunology, 23*(3), 125–132. https://doi.org/10.1016/j.trim.2010.05.003.

Ostalska-Nowicka, D., Nowicki, M., Zachwieja, J., Kasper, M., & Witt, M. (2007). The significance of caveolin-1 expression in parietal epithelial cells of Bowman's capsule. *Histopathology, 51*(5), 611–621. https://doi.org/10.1111/j.1365-2559.2007.02844.x.

Park, C. H., & Maack, T. (1984). Albumin absorption and catabolism by isolated perfused proximal convoluted tubules of the rabbit. *Journal of Clinical Investigation, 73*(3), 767–777. https://doi.org/10.1172/JCI111270.

Pearson, A. L., Colville-Nash, P., Kwan, J. T. C., & Dockrell, M. E. C. (2008). Albumin induces interleukin-6 release from primary human proximal tubule epithelial cells. *Journal of Nephrology, 21*(6), 887–893.

Peres, R. A. S., Peruchetti, D. B., Silva-Aguiar, R. P., Teixeira, D. E., Gomes, C. P., Takiya, C. M., ... Caruso-Neves, C. (2023). Rapamycin treatment induces tubular proteinuria: Role of megalin-mediated protein reabsorption. *Frontiers in Pharmacology, 14*, 1194816. https://doi.org/10.3389/fphar.2023.1194816.

Peruchetti, D. B., Barahuna-Filho, P. F. R., Silva-Aguiar, R. P., Abreu, T. P., Takiya, C. M., Cheng, J., ... Caruso-Neves, C. (2021). Megalin-mediated albumin endocytosis in renal proximal tubules is involved in the antiproteinuric effect of angiotensin II type 1 receptor blocker in a subclinical acute kidney injury animal model. *Biochimica Et Biophysica Acta. General Subjects, 1865*(9), 129950. https://doi.org/10.1016/j.bbagen.2021.129950.

Peruchetti, D. B., Cheng, J., Caruso-Neves, C., & Guggino, W. B. (2014). Mis-regulation of mammalian target of rapamycin (mTOR) complexes induced by albuminuria in proximal tubules. *The Journal of Biological Chemistry, 289*(24), 16790–16801. https://doi.org/10.1074/jbc.M114.549717.

Peruchetti, D. B., Silva-Filho, J. L., Silva-Aguiar, R. P., Teixeira, D. E., Takiya, C. M., Souza, M. C., ... Caruso-Neves, C. (2020). IL-4 receptor α chain protects the kidney against tubule-interstitial injury induced by albumin overload. *Frontiers in Physiology, 11*, 172. https://doi.org/10.3389/fphys.2020.00172.

Peruchetti, D. de B., Silva-Aguiar, R. P., Siqueira, G. M., Dias, W. B., & Caruso-Neves, C. (2018). High glucose reduces megalin-mediated albumin endocytosis in renal proximal tubule cells through protein kinase B O-GlcNAcylation. *The Journal of Biological Chemistry, 293*(29), 11388–11400. https://doi.org/10.1074/jbc.RA117.001337.

Polesel, M., & Hall, A. M. (2019). Axial differences in endocytosis along the kidney proximal tubule. *American Journal of Physiology. Renal Physiology, 317*(6), F1526–F1530. https://doi.org/10.1152/ajprenal.00459.2019.

Raghavan, V., Rbaibi, Y., Pastor-Soler, N. M., Carattino, M. D., & Weisz, O. A. (2014). Shear stress-dependent regulation of apical endocytosis in renal proximal tubule cells mediated by primary cilia. *Proceedings of the National Academy of Sciences of the United States of America, 111*(23), 8506–8511. https://doi.org/10.1073/pnas.1402195111.

Rayner, H. C., Horsburgh, T., Brown, S. L., Lavender, L., Winder, A. F., & Walls, J. (1990). Receptor-mediated endocytosis of low-density lipoprotein by cultured human glomerular cells. *Nephron, 55*(3), 292–299. https://doi.org/10.1159/000185978.

Rennick, J. J., Johnston, A. P. R., & Parton, R. G. (2021). Key principles and methods for studying the endocytosis of biological and nanoparticle therapeutics. *Nature Nanotechnology, 16*(3), 266–276. https://doi.org/10.1038/s41565-021-00858-8.

Runyan, C. E., Schnaper, H. W., & Poncelet, A.-C. (2005). The role of internalization in transforming growth factor β1-induced Smad2 association with smad anchor for receptor activation (SARA) and Smad2-dependent signaling in human mesangial cells. *Journal of Biological Chemistry, 280*(9), 8300–8308. https://doi.org/10.1074/jbc.M407939200.

Russo, L. M., Sandoval, R. M., McKee, M., Osicka, T. M., Collins, A. B., Brown, D., ... Comper, W. D. (2007). The normal kidney filters nephrotic levels of albumin retrieved by proximal tubule cells: Retrieval is disrupted in nephrotic states. *Kidney International, 71*(6), 504–513. https://doi.org/10.1038/sj.ki.5002041.

Schießl, I. M., Hammer, A., Kattler, V., Gess, B., Theilig, F., Witzgall, R., & Castrop, H. (2016). Intravital imaging reveals angiotensin II-induced transcytosis of albumin by podocytes. *Journal of the American Society of Nephrology: JASN, 27*(3), 731–744. https://doi.org/10.1681/ASN.2014111125.

Schwegler, J. S., Heppelmann, B., Mildenberger, S., & Silbernagl, S. (1991). Receptor-mediated endocytosis of albumin in cultured opossum kidney cells: A model for proximal tubular protein reabsorption. *Pflugers Archiv: European Journal of Physiology, 418*(4), 383–392. https://doi.org/10.1007/BF00550876.

Shinkai, Y. (1982). Experimental glomerulonephritis induced in rabbits by horseradish peroxidase. Mesangial uptake and processing of immune complexes. *Laboratory Investigation; a Journal of Technical Methods and Pathology, 46*(6), 577–583.

Silva-Aguiar, R. P., Peruchetti, D. B., Florentino, L. S., Takiya, C. M., Marzolo, M.-P., Dias, W. B., ... Caruso-Neves, C. (2022). Albumin expands albumin reabsorption capacity in proximal tubule epithelial cells through a positive feedback loop between AKT and megalin. *International Journal of Molecular Sciences, 23*(2), 848. https://doi.org/10.3390/ijms23020848.

Silva-Aguiar, R. P., Teixeira, D. E., Peruchetti, D. B., Florentino, L. S., Peres, R. A. S., Gomes, C. P., ... Caruso-Neves, C. (2022). SARS-CoV-2 spike protein inhibits megalin-mediated albumin endocytosis in proximal tubule epithelial cells. *Biochimica Et Biophysica Acta. Molecular Basis of Disease, 1868*(12), 166496. https://doi.org/10.1016/j.bbadis.2022.166496.

Singhal, P. C., Ding, G. H., DeCandido, S., Franki, N., Hays, R. M., & Schlondorff, D. (1987). Endocytosis by cultured mesangial cells and associated changes in prostaglandin E2 synthesis. *The American Journal of Physiology, 252*(4 Pt 2), F627–F634. https://doi.org/10.1152/ajprenal.1987.252.4.F627.

Singhal, P. C., Franki, N., Gibbons, N., & Hays, R. M. (1992). Effects of angiotensin II and arginine vasopressin on F-actin content of cultured mesangial cells. *Journal of the American Society of Nephrology, 3*(1), 80–87. https://doi.org/10.1681/ASN.V3180.

Singhal, P. C., Gupta, S., Sharma, P., Shah, H., Shah, N., & Patel, P. (2000). No title found. *Inflammation, 24*(6), 519–532. https://doi.org/10.1023/A:1007073306394.

Takakura, Y., Morita, T., Fujikawa, M., Hayashi, M., Sezaki, H., Hashida, M., & Borchardt, R. T. (1995). Characterization of LLC-PK1 kidney epithelial cells as an in vitro model for studying renal tubular reabsorption of protein drugs. *Pharmaceutical Research, 12*(12), 1968–1972. https://doi.org/10.1023/a:1016256325921.

Takano, M., Nakanishi, N., Kitahara, Y., Sasaki, Y., Murakami, T., & Nagai, J. (2002). Cisplatin-induced inhibition of receptor-mediated endocytosis of protein in the kidney. *Kidney International, 62*(5), 1707–1717. https://doi.org/10.1046/j.1523-1755.2002.00623.x.

Tang, S., Leung, J. C. K., Abe, K., Chan, K. W., Chan, L. Y. Y., Chan, T. M., & Lai, K. N. (2003). Albumin stimulates interleukin-8 expression in proximal tubular epithelial cells in vitro and in vivo. *The Journal of Clinical Investigation, 111*(4), 515–527. https://doi.org/10.1172/JCI16079.

Teixeira, D. E., Peruchetti, D. B., Silva, L. S., Silva-Aguiar, R. P., Oquendo, M. B., Silva-Filho, J. L., ... Caruso-Neves, C. (2019). Lithium ameliorates tubule-interstitial injury through activation of the mTORC2/protein kinase B pathway. *PLoS One, 14*(4), e0215871. https://doi.org/10.1371/journal.pone.0215871.

Teng, J., Russell, W. J., Gu, X., Cardelli, J., Jones, M. L., & Herrera, G. A. (2004). Different types of glomerulopathic light chains interact with mesangial cells using a common receptor but exhibit different intracellular trafficking patterns. *Laboratory Investigation, 84*(4), 440–451. https://doi.org/10.1038/labinvest.3700069.

Tenten, V., Menzel, S., Kunter, U., Sicking, E.-M., van Roeyen, C. R. C., Sanden, S. K., ... Moeller, M. J. (2013). Albumin is recycled from the primary urine by tubular transcytosis. *Journal of the American Society of Nephrology: JASN, 24*(12), 1966–1980. https://doi.org/10.1681/ASN.2013010018.

Terryn, S., Tanaka, K., Lengelé, J.-P., Olinger, E., Dubois-Laforgue, D., Garbay, S., ... Devuyst, O. (2016). Tubular proteinuria in patients with HNF1α mutations: HNF1α drives endocytosis in the proximal tubule. *Kidney International, 89*(5), 1075–1089. https://doi.org/10.1016/j.kint.2016.01.027.

Thomas, M. E., Brunskill, N. J., Harris, K. P., Bailey, E., Pringle, J. H., Furness, P. N., & Walls, J. (1999). Proteinuria induces tubular cell turnover: A potential mechanism for tubular atrophy. *Kidney International, 55*(3), 890–898. https://doi.org/10.1046/j.1523-1755.1999.055003890.x.

Webber, W. A., & Wong, W. T. (1973). The function of the basal filaments in the parietal layer of Bowman's capsule. *Canadian Journal of Physiology and Pharmacology, 51*(2), 53–60. https://doi.org/10.1139/y73-008.

Werion, A., Belkhir, L., Perrot, M., Schmit, G., Aydin, S., Chen, Z., ... Jadoul, M. Cliniques universitaires Saint-Luc (CUSL) COVID-19 Research Group. (2020). SARS-CoV-2 causes a specific dysfunction of the kidney proximal tubule. *Kidney International, 98*(5), 1296–1307. https://doi.org/10.1016/j.kint.2020.07.019.

Wheeler, D. C., Fernando, R. L., Gillett, M. P. T., Zaruba, J., Persaud, J., Kingstone, D., ... Moorhead, J. F. (1991). Characterisation of the binding of low-density lipoproteins to cultured rat mesangial cells. *Nephrology Dialysis Transplantation, 6*(10), 701–708. https://doi.org/10.1093/ndt/6.10.701.

Wohlfarth, V., Drumm, K., Mildenberger, S., Freudinger, R., & Gekle, M. (2003). Protein uptake disturbs collagen homeostasis in proximal tubule-derived cells. *Kidney International. Supplement, 84*, S103–S109. https://doi.org/10.1046/j.1523-1755.63.s84.13.x.

Wu, D., Yang, X., Zheng, T., Xing, S., Wang, J., Chi, J., ... Jin, S. (2016). A novel mechanism of action for salidroside to alleviate diabetic albuminuria: Effects on albumin transcytosis across glomerular endothelial cells. *American Journal of Physiology-Endocrinology and Metabolism, 310*(3), E225–E237. https://doi.org/10.1152/ajpendo.00391.2015.

Yang, Q., Ma, Y., Liu, Y., Liang, W., Chen, X., Ren, Z., ... Ding, G. (2016). Angiotensin II down-regulates nephrin–Akt signaling and induces podocyte injury: Role of c-Abl. *Molecular Biology of the Cell, 27*(1), 197–208. https://doi.org/10.1091/mbc.E15-04-0223.

Yoder, B. K. (2007). Role of primary cilia in the pathogenesis of polycystic kidney disease. *Journal of the American Society of Nephrology, 18*(5), 1381–1388. https://doi.org/10.1681/ASN.2006111215.

Zeng, B., Chen, G.-L., Garcia-Vaz, E., Bhandari, S., Daskoulidou, N., Berglund, L. M., ... Xu, S.-Z. (2017). ORAI channels are critical for receptor-mediated endocytosis of albumin. *Nature Communications, 8*(1), 1920. https://doi.org/10.1038/s41467-017-02094-y.

Zhai, X. Y., Nielsen, R., Birn, H., Drumm, K., Mildenberger, S., Freudinger, R., ... Gekle, M. (2000). Cubilin- and megalin-mediated uptake of albumin in cultured proximal tubule cells of opossum kidney. *Kidney International, 58*(4), 1523–1533. https://doi.org/10.1046/j.1523-1755.2000.00314.x.

Zhao, X., Chen, X., Chima, A., Zhang, Y., George, J., Cobbs, A., & Emmett, N. (2019). Albumin induces CD44 expression in glomerular parietal epithelial cells by activating extracellular signal-regulated kinase 1/2 pathway. *Journal of Cellular Physiology, 234*(5), 7224–7235. https://doi.org/10.1002/jcp.27477.

Zhao, X., Jiang, C., Olufade, R., Liu, D., & Emmett, N. (2016). Kidney injury molecule-1 enhances endocytosis of albumin in renal proximal tubular cells. *Journal of Cellular Physiology, 231*(4), 896–907. https://doi.org/10.1002/jcp.25181.

CHAPTER TWO

Endocytosis in malaria parasites: An ultrastructural perspective of membrane interplay in a unique infection model

Camila Wendt[a,b,*] and Kildare Miranda[a,*]
[a]Laboratório de Ultraestrutura Celular Hertha Meyer, Instituto de Biofísica Carlos Chagas Filho and Centro Nacional de Biologia Estrutural e Bioimagem, Universidade Federal do Rio de Janeiro, Rio de Janeiro, Brazil
[b]Laboratório de Biomineralização, Instituto de Ciências Biomédicas, Universidade Federal do Rio de Janeiro, Rio de Janeiro, Brazil
*Corresponding authors. e-mail address: camilawendt@biof.ufrj.br; kmiranda@biof.ufrj.br

Contents

Abstract

Malaria remains a major global threat, representing a severe public health problem worldwide. Annually, it is responsible for a high rate of morbidity and mortality in many tropical developing countries where the disease is endemic. The causative agent of malaria, *Plasmodium* spp., exhibits a complex life cycle, alternating between an invertebrate vector, which transmits the disease, and the vertebrate host. The disease pathology observed in the vertebrate host is attributed to the asexual development of *Plasmodium* spp. inside the erythrocyte. Once inside the red blood cell, malaria parasites cause extensive changes in the host cell, increasing membrane rigidity and altering its normal discoid shape. Additionally, during their intraerythrocytic development, malaria parasites incorporate and degrade up to 70 % of host cell hemoglobin. This mechanism is essential for parasite development and represents an important drug target. Blocking the steps related to hemoglobin endocytosis or degradation impairs parasite development and can lead to its death.

Current Topics in Membranes, Volume 93
ISSN 1063-5823, https://doi.org/10.1016/bs.ctm.2024.05.001

The ultrastructural analysis of hemoglobin endocytosis on *Plasmodium* spp. has been broadly explored along the years. However, it is only recently that the proteins involved in this process have started to emerge. Here, we will review the most important features related to hemoglobin endocytosis and catabolism on malaria parasites. A special focus will be given to the recent analysis obtained through 3D visualization approaches and to the molecules involved in these mechanisms.

1. Plasmodium spp. life cycle

The fight against malaria is a global effort. Although rigorous control programs have achieved substantial reductions in the global malaria burden, the number of malaria cases and deaths in 2022 was significantly higher than before the pandemic in 2019. Most deaths are attributed to *Plasmodium falciparum*, one of the five species that infect humans. However, higher rates of morbidity have also been associated to *P. vivax*, the human malaria species that has the largest global distribution (WHO, 2023).

Malaria parasites exhibit a complex life cycle, alternating between an invertebrate and a vertebrate host (Fig. 1). Disease transmission occurs through the bite of the mosquito vector. During the blood meal, the sporozoites present on *Anopheles* spp. mosquito salivary glands are injected into the vertebrated vector blood vessels (Ménard et al., 2013; Sinnis & Zavala, 2008, 2012). Sporozoites have a high mobility and migrate to the liver, invading the hepatocytes and promoting multiple cycles of nuclear replication, generating a higher quantity of merozoites. After being released from hepatocytes, the merozoites enter the blood vessels and invade the erythrocytes. The liver stage development is asymptomatic, whereas all the clinical symptoms associated to malaria disease are evident in the blood asexual development (Prudêncio, Rodriguez, & Mota, 2006; Silvie, Mota, Matuschewski, & Prudêncio, 2008). During the erythrocytic cycle, the parasite develops through the stages of ring, trophozoite and schizont. The late schizont stage is characterized by the presence of multiple merozoites soon to be released. Once mature, the merozoites promote blood cell rupture, being released in the bloodstream, where they infect other erythrocytes (Bannister, Hopkins, Fowler, Krishna, & Mitchell, 2000). A detailed morphological description of the intraerythrocytic development stage of malaria parasites has been recently reviewed by Miranda, Wendt, Gomes, and de Souza (2022). Although cell cycle duration varies across different malaria parasite species, it generally occurs in multiples of 24 h. *P. falciparum* and *P. vivax*, for

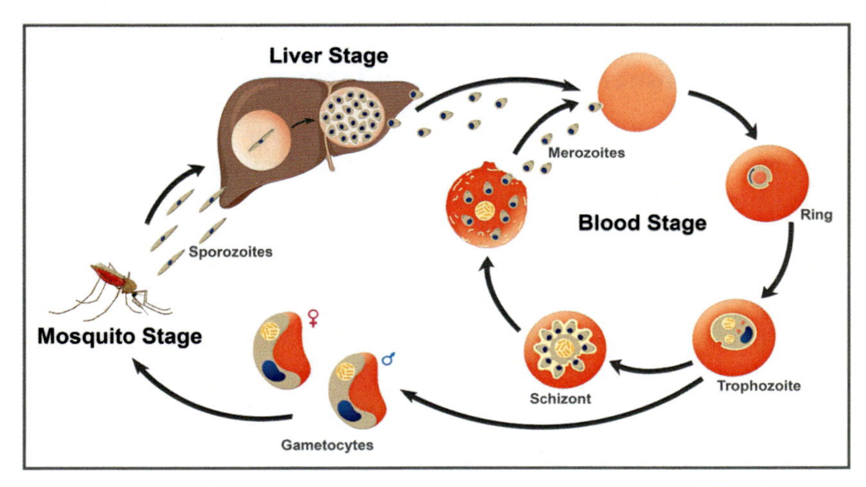

Fig. 1 *Plasmodium* spp. development inside the vertebrate host. Sporozoite stage parasites are injected into the vertebrate host dermis by a feeding female anopheline mosquito. Sporozoites actively reach the circulation and migrate to the liver, invading the hepatocytes. Following development in the liver, the sporozoite undergoes replication, originating liver stage merozoites. These parasites are released from the hepatocytes into the blood where they invade erythrocytes and start the asexual replication development (blood stage). Inside the erythrocyte, the parasite develops through the stages of ring, trophozoite and schizont; replicating to produce merozoites that are released during the egress. Free merozoites invade non-infected erythrocytes, starting a new asexual blood stage cycle. In each erythrocytic cycle, a proportion of trophozoites undergo gametocytogenesis, originating male and female gametocytes. Mature gametocytes remain in the bloodstream, being taken up by the mosquito during feeding. Sexual development occurs inside the insect, generating new sporozoites. These will migrate to the salivary glands, being injected in the vertebrate host skin during the next blood meal.

instance, possess an intraerythrocytic development of 48 h (Cowman, Healer, Marapana, & Marsh, 2016; De Niz & Heussler, 2018). Alternatively, instead of differentiating into schizont stages, a small proportion of parasites undergo a developmental switch, originating male and female gametocytes. The mechanisms that trigger this transition are still under discussion, however it has been suggested that a stimulus, such as high parasitemia or exposure to antimalarial drugs cause an increase in the conversion rate to gametocytes. Ingestion of the gametocytes by the mosquito vector will initiate *Plasmodium* spp. sexual development, culminating in the generation of new sporozoite stage parasites (Josling & Llinás, 2015).

2. Development inside the erythrocyte: a unique environment of membrane interplay

Red blood cell invasion by *Plasmodium* spp. parasites is coordinated by the merozoite stage. The invasion occurs through the formation of a tight junction between the parasite and the host membrane. Powered by its actin-myosin motor, the parasite propels itself into the erythrocyte, creating a parasitophorous vacuole (Cowman & Crabb, 2006; Cowman, Berry, & Baum, 2012). The parasitophorous vacuolar membrane (PVM) is closely apposed to the parasite plasma membrane (PPM). Attachment points between the two membranes were observed, suggesting that these regions might be related to exchange of lipids across the parasitophorous vacuole space (Garten et al., 2019).

The composition of the parasitophorous vacuole is still debated. Indeed, as the red blood cell surface area remains constant before and after invasion, it is accepted that there is a substantial contribution of parasite-derived phospholipids to parasitophorous vacuole biogenesis. During invasion, additional proteins located at the parasite apical organelles are released and incorporated into the parasitophorous vacuole membrane (Goldberg & Zimmerberg, 2020; Lauer et al., 2000; Murphy et al., 2006; Ward, Miller, & Dvorak, 1993).

Developing inside a parasitophorous vacuole in a highly differentiated cell, such as the erythrocyte, imposes some challenges to the parasite. It depends on a vast range of essential nutrients that cannot be harvested from the host cell, as human erythrocytes are devoid of organelles, lack de novo protein synthesis and endogenous protein trafficking machinery. In this regard, specific mechanisms related to protein export to the host cell surface and nutrient uptake pathways are observed shortly after merozoite invasion (Boddey & Cowman, 2013).

The erythrocyte plasma membrane has an endogenous range of specialized transporters (channels, carriers, and pumps) to facilitate the passage of substrates. Nevertheless, not all substrates required for *Plasmodium* spp. development, such as isoleucine, can be obtained via host cell transporters, or, in other cases, they are transported in insufficient amount (Elliott, Saliba, & Kirk, 2001; Martin & Kirk, 2007; Saliba, Horner, & Kirk, 1998). Analysis of *P. falciparum* infected erythrocytes showed that the host cell exhibits an increase in the permeability to a diverse range of solutes, such as sugars, amino acids, vitamins, and ions. Such changes in permeability are attributed to the establishment of new permeation pathways (NPPs). The

NPP acts as a conduit for the uptake of nutrients and exchange of waste products, allowing parasites to condition the intracellular ionic environment (Martin, 2020).

Another key feature for *Plasmodium* spp. developing inside the erythrocyte is related to the mechanism of protein export via an exomembrane system and host cell remodeling. Different structures were attributed to the exomembrane system. These includes the tubovesicular network (TVN), Maurer's clefts (MCs), caveola-vesicle complex (CVC), J dots and small vesicles. The complexity of the exomembrane system varies among *Plasmodium* species. Maurer's clefts, for instance, represent a characteristic structure described in *P. falciparum* whereas the TVN is broadly observed in human and murine malaria parasites (Frischknecht & Lanzer, 2008; Sherling & van Ooij, 2016; Soares-Medeiros, de Souza, Jiao, Barrabin, & Miranda, 2012). Even though morphological similarities are observed in the exomembrane system among different malaria species, further studies are necessary to establish the molecular mechanisms underlying their exact functional roles.

During its intraerythrocytic development, malaria parasites also absorb large amounts of the host cell cytoplasm via endocytosis. This represents an important mechanism through which the parasite carries out the uptake and catabolism of up to 70% of host cell cytoplasm (Francis, Sullivan, & Goldberg, 1997; Krugliak, Zhang, & Ginsburg, 2002). The internalized hemoglobin is later metabolized in the food vacuole, an acid organelle that has specific proteases that will promote hemoglobin catabolism. Hemoglobin proteolysis releases heme and generates amino acids that are transported to the parasite cytoplasm whereas free heme remains within the food vacuole. Free, dissociated heme, can induce a variety of cytotoxic effects, such as inhibition of protein function, disruption of membrane bilayers, and produce a diverse range of reactive oxygen species (Ponka, 1999). Such effects are minimized with the immobilization of free heme in the form of hemozoin crystals. This mechanism is essential to parasite survival inside the red blood cell and represents so far, the main target for the drugs currently used on malaria therapy (Sigala & Goldberg, 2014).

Interestingly, hemoglobin digestion provides far more amino acids than the required for protein biosynthesis during the intraerythrocytic development. The only exception is isoleucine, which is not present in adult human hemoglobin and must be obtained from the extracellular medium (Liu, Istvan, Gluzman, Gross, & Goldberg, 2006). The redundant amino acid incorporation via hemoglobin endocytosis and catabolism represented for many years an unresolved puzzle on malaria research. The prevailing belief that hemoglobin degradation is essential for parasite survival is based on analyses using protease inhibitors. The blockage of hemoglobin degradation impairs parasite development (Goldberg, 1990; Liu et al., 2006). Based on such studies, the current understanding converges on the idea that the energetically costly mechanism of hemoglobin digestion represents an adaptation for reducing colloid-osmotic pressure, thereby avoiding premature lysis of the host cell and creating free space for parasite growth. (Hanssen et al., 2012; Lew, Tiffert, & Ginsburg, 2003; Mauritz et al., 2009; Staines, Ellory, & Kirk, 2001).

3. Hemoglobin uptake: a membrane-focused structural analysis

Groundbreaking works using transmission electron microscopy (TEM) showed that hemoglobin uptake occurs through cytostomes (Aikawa, Hepler, Huff, & Sprinz, 1966; Aikawa, Huff, & Sprinz, 1969b; Aikawa, 1971). In *Plasmodium* spp. the cytostome is formed through the synchronous invagination of the parasite plasma membrane (PPM) and the parasitophorous vacuole membrane (PVM) (Fig. 2). TEM data showed cytostomes as pear shaped structures presenting an electron dense collar-decorated opening that remains constant in terms of size and shape. As hemoglobin uptake progresses, the cytostome tube enlarges. At this stage, the electron density of the cytostome lumen is identical to the red cell cytoplasm. Later, hemoglobin filled vacuoles are pinched off from the cytostomal tube and directed to the food vacuole, where hemoglobin catabolism occurs. The food vacuole has only a single membrane whereas the cytostome has a double membrane (one from the PVM and other from the PPM) (Aikawa et al., 1966).

Early analysis of hemoglobin uptake on malaria parasites were manly achieved by the TEM analysis of avian and rodent malaria models (Aikawa et al., 1966; Rudzinska & Trager, 1957, 1959; Slomianny, Prensier, & Charet, 1985). Later, human malaria models, especially *P. falciparum*, became an important model to study hemoglobin endocytosis on malaria parasites (Abu Bakar, Klonis, Hanssen, Chan, & Tilley, 2010; Elliott et al., 2008; Lazarus, Schneider, & Taraschi, 2008; Milani, Schneider, & Taraschi, 2015). The name of the structure involved on hemoglobin uptake, the cytostome, was given because of its involvement in endocytosis of host cell

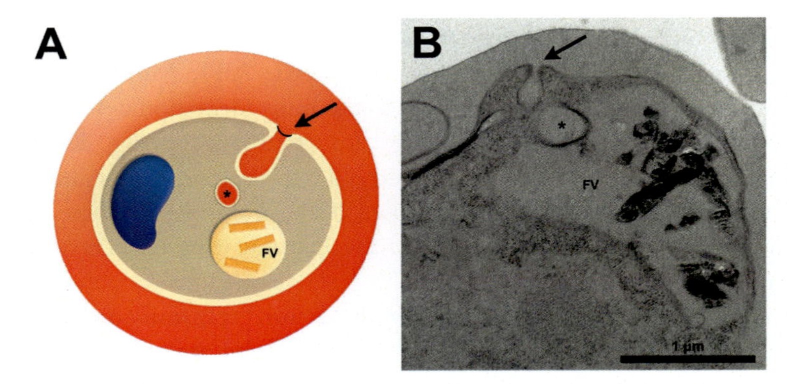

Fig. 2 Hemoglobin endocytosis pathway proposed by Aikawa et al. (1966). (A) Endocytosis of host cell cytosol occurs via cytostomes (arrow). This structure originates by the invagination of the parasitophorous vacuole and the parasite plasma membrane, forming a tube that propagates throughout the parasite cytosol. A characteristic electron dense ring is observed close to the cytostome aperture. Hemoglobin filled vacuoles (asterisk) buds from the end portion of the cytostomal tube and are delivered to the food vacuole (FV), where hemoglobin catabolism occurs. (B) TEM image of *P. falciparum* trophozoite stage showing a cytostome structure (arrow), a hemoglobin filled vacuole (asterisk) and the food vacuole (FV). The cytostome tube and hemoglobin vacuole has a content that has the same electron density of the host cell cytoplasm whereas more electron dense structures (hemozoin crystals) are observed inside the food vacuole.

cytosol. The term cytostome has been initially used at the early light microscope studies as a generic word to describe the pore through which ciliate and flagellate protozoa engulfed food (Aikawa et al., 1966). Later, it has also been used to describe a structure used by trypanosomatids to obtain macromolecules from the extracellular milleu (Preston, 1969; Steinert & Novikoff, 1960). Although in all cases the cytostome is a structure involved in the uptake of extracellular content, these structures are not morphologically related. The cytostome in trypanosomatids consists of a continuous opening at the cell surface, that forms a profound invagination, known as cytopharynx, towards the cell cytoplasm. The cytopharynx is sustained by a set of microtubules, that accompany the cytopharynx along its entire length (Alcantara, De, Vidal, De Souza, & Cunha-e-Silva, 2014; Preston, 1969). In *Plasmodium* spp. the cytostome is a transient structure which can originate at any site of the parasite membrane and so far has not been shown to be supported by microtubules.

Since its early description of the cytostome on malaria parasites in 1966 (Aikawa et al., 1966), electron microscopy techniques have evolved, developing tools that allow the 3D characterization of different structures (Eshar, Dahan-Pasternak, Weiner, & Dzikowski, 2011; Miranda, Girard-dias, Attias, & Souza, 2015). Associated to the instrumentation development, sample preservation techniques were also improved, being developed cryopreservation methods which allowed cell ultrastructure preservation in a close to native state (Hurbain & Sachse, 2011). Considering these recent technical advances, hemoglobin uptake mechanisms have been revisited in the past decade.

Analysis of *P. falciparum* by serial section transmission electron microscopy provided interesting data on hemoglobin uptake mechanisms during asexual development, pointing out the existence of specific hemoglobin uptake mechanisms at each developmental stage. It was shown that, after erythrocyte invasion, the parasites undergo morphological transformations in which they fold, like a cup, onto themselves, promoting the uptake of a large portion of host cell cytoplasm (Fig. 3). This mechanism was termed as the 'big gulp', since large amounts of hemoglobin are endocytosed, representing about 40% of total parasite volume (Elliott et al., 2008). Although well documented, it is still discussed if the 'big gulp theory' indeed represented the only hemoglobin uptake mechanism at early stages (Abu Bakar et al., 2010; Hanssen et al., 2011). Different studies have already described the ultrastructure of the ring stage parasite, pointing that after erythrocyte invasion the parasite developed a cup shape morphology (Bannister et al., 2000; Bannister, Hopkins, Margos, Dluzewski, & Mitchell, 2004; Langreth, Jensen, Reese, & Trager, 1978). However, it was suggested that the invagination formed during the "big gulp" does not form a sealed intra-parasite compartment. It has been discussed that it might simply reflect the need to increase surface area without a significant increase in volume (Abu Bakar et al., 2010; Bannister et al., 2004; Hanssen et al., 2011). Nevertheless, electron tomography analysis of high pressure frozen *P. chabaudi* ring stage parasites revealed a large hemoglobin containing vacuole that had no connection to the host cell cytoplasm, corroborating the big-gulp theory (Wendt, Rachid, Souza, De, & Miranda, 2016). Besides the big gulp, other hemoglobin uptake mechanisms must be active at the ring stage parasite, as multiple food vacuoles are already observed at this developmental stage (Bannister et al., 2004; Wendt et al., 2016). So far, there has been a TEM description of a single active cytostome on *P. falciparum* ring stage, however, this has not been explored on further works (Bannister et al., 2004).

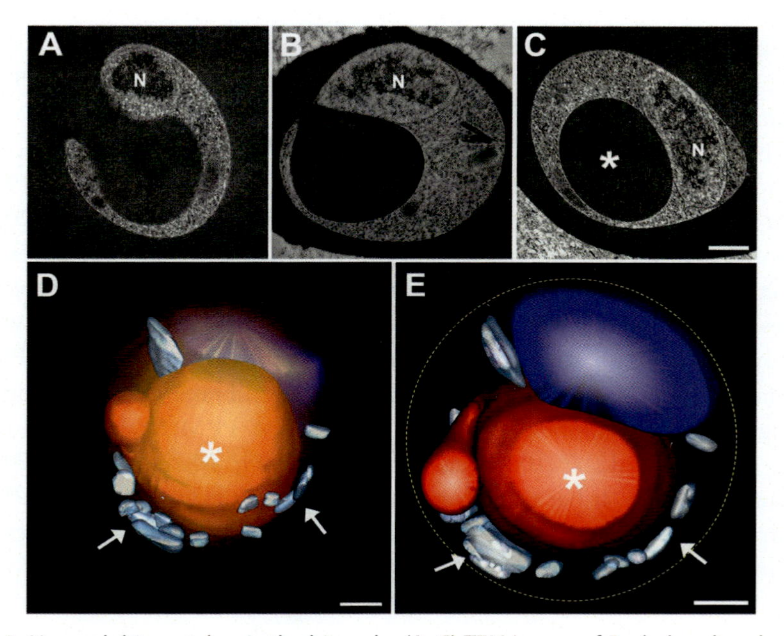

Fig. 3 Hemoglobin uptake via the big gulp. (A–C) TEM image of *P. chabaudi* early ring stages showing the mechanism of adopting a cup shape parasite and expanding their edge until forming a large vacuole filled with hemoglobin (asterisk). (D and E) 3D rendering of a ring stage showing that most of cell volume is occupied by a large hemoglobin vacuole (asterisk) whereas small food vacuoles (arrows) are observed in the periphery of the parasite cytoplasm. N: nucleus. Scale bar: 500 nm. *Adapted from Wendt, C., Rachid, R., De Souza, W., Miranda, K. (2016). Electron tomography characterization of hemoglobin uptake in Plasmodium chabaudi reveals a stage-dependent mechanism for food vacuole morphogenesis. Journal of Structural Biology, 194, 171–179. https://doi.org/10.1016/j.jsb.2016.02.014.*

Most of the data regarding hemoglobin endocytosis on malaria parasites were obtained from the analysis of trophozoites. At this stage, parasite volume increases considerably, a phenomenon that is accompanied by the expansion of key organelles needed for metabolism (e.g. nucleus, endomembrane system, mitochondria, and apicoplast) (Aikawa, 1971; Bannister et al., 2000). However, slight volume changes occur within the infected erythrocyte, suggesting that the increase in parasite volume is coupled to a massive consumption of the host cell content (Hanssen et al., 2012; Mauritz et al., 2011). Hemoglobin uptake at the trophozoite stage occurs via cytostomes (Abu Aikawa et al., 1966; Bakar et al., 2010; Elliott et al., 2008; Lazarus et al., 2008; Liu et al., 2019;

Milani et al., 2015; Slomianny et al., 1985; Slomianny, 1990; Wendt et al., 2016) and phagotrophs (Abu Bakar et al., 2010; Elliott et al., 2008; Milani et al., 2015).

The cytostome, in its classical description, has been observed in the trophozoite stage of different *Plasmodium* species (Abu Aikawa et al., 1966; Bakar et al., 2010; Elliott et al., 2008; Lazarus et al., 2008; Liu et al., 2019; Milani et al., 2015; Slomianny et al., 1985; Slomianny, 1990; Wendt et al., 2016). Electron tomography indicated an average of two to four cytostomes in each parasite. It was also reported that these structures were usually located in clusters at specific regions along the parasite PVM/PPM interface, which may be a result of favorable membrane conditions originating at specific membrane domains in the PVM/PPM, such as lipid rafts (Abu Bakar et al., 2010; Lazarus et al., 2008; Milani et al., 2015; Slomianny, 1990). It has been suggested that lipid rafts localize to the PVM (Murphy et al., 2004), and may function as initiation sites for endocytosis (Helms & Zurzolo, 2004). In contrast to *P. falciparum*, larger cytostomal tubes as well as hemoglobin containing structures have been described in malaria murine model *P. chabaudi* (Slomianny et al., 1985; Soares-Medeiros et al., 2012; Wendt et al., 2016) and in the simian/human malaria specie, *P. knowlesi* (Liu et al., 2019).

Besides the cytostome, another structure, termed as phagotroph, is also related to hemoglobin uptake at the late trophozoite stage. Phagotrophs are host cell cytosol-filled invaginations that have an aperture larger than cytostomes. The distinctive electron dense collar is not observed on these structures and its opening is wider than the 200–300 nm usually observed in cytostomes. Additionally, the phagotroph originates a larger hemoglobin filled structure when compared to the cytostome tube. Another distinctive feature of this structure is that, unlike the cytostome, phagotroph formation is actin independent (Elliott et al., 2008; Milani et al., 2015).

So far, few works have attempted to explore hemoglobin uptake mechanisms at the schizont and gametocytes stages. Observation of *P. falciparum* asexual development stages by soft x-ray tomography indicated that, even reduced, hemoglobin degradation continues during late schizont and late gametocytes stages (Hanssen et al., 2011). This analysis raises a question if this hemoglobin degradation is associated to a hemoglobin uptake pathway or to degradation of reminiscent hemoglobin that has been previously internalized. Nevertheless, active hemoglobin uptake by cytostomes has been described in *Plasmodium* spp. schizont stage (Abu Aikawa et al., 1966; Aikawa, 1971; Bakar et al., 2010; Wendt et al., 2016). The schizont is the last stage observed during *Plasmodium* spp. intraerythrocytic development. Repetitive nuclear division starts in early schizonts until individual merozoites are observed at the late stage. TEM analysis showed that these newly formed merozoites have a cytostome in the region between the cell anterior and posterior end (Aikawa & Jordan, 1968; Aikawa et al., 1966; Aikawa, Huff, & Sprinz, 1967; Aikawa, Cook, Sakoda, & Sprinz, 1969a; Bannister et al., 2003; Ladda, 1969; Scalzi & Bahr, 1968). However, in contrast to the other erythrocytic stages, the cytostomal cavity observed on the merozoite is electron-transparent and is not functional (Aikawa et al., 1966; Aikawa, 1971). It is possible that this structure might become active after erythrocyte invasion during the ring stage development.

Little is known about hemoglobin uptake at the gametocyte stage. Gametocyte differentiation includes several different stages, which varies along the malaria species. *P. falciparum* gametocytes, for example, has five distinct gametocyte developmental stages, whereas other humans, rodent and avian malaria parasites have one or two (Dixon, Dearnley, Hanssen, Gilberger, & Tilley, 2012; Ngotho et al., 2019). TEM observations of *P. gallinaceum* (Aikawa, Ito, & Nijhout, 1984) and *P. falciparum* (Kass, Willerson, Rieckmann, Carson, & Becker, 1971; Ponnudurai, Lensen, Meis, & Meuwissen, 1986) gametocytes showed that hemoglobin uptake at this stage occurs via cytostomes.

4. Molecular basis of hemoglobin uptake mechanisms

The molecules involved on hemoglobin endocytosis pathway are still under debate. Inhibitor studies suggested a role of actin on endocytosis (Elliott et al., 2008; Lazarus et al., 2008; Milani et al., 2015; Smythe, Joiner, & Hoppe, 2008). The majority of actin in apicomplexan parasites exists in a soluble form, presumably in a monomeric state and, unlike what is reported in most eukaryotes, apicomplexan actin forms only short filaments of less than 100 nm (Olshina, Wong, & Baum, 2012; Pospich et al., 2017). It was shown that actin stabilization disrupts cytostome morphology and inhibits hemoglobin transport whereas actin filaments destabilization increases the number hemoglobin filled structures (Lazarus et al., 2008; Milani et al., 2015). Nevertheless, actin dynamics does not interfere on hemoglobin uptake via the big gulp or phagotroph (Elliott et al., 2008; Milani et al., 2015).

The molecular composition of the characteristic electron dense collar observed on the cytostome aperture is still unknown. Since its initial description, it has been assumed that the purpose of the cytostomal collar was to aid the fission of the cytostome from the PVM-PPM interface, originating a hemoglobin filled structure (Aikawa et al., 1966). Therefore, there was a discussion that the dynamin might play a role on the cytostome fission mechanism. Milani et al. attempted to use immuno-EM to localize dynamin in trophozoites of *P. falciparum* (Milani et al., 2015). While dynamin was absent from the collar, it appeared to localize to membranous structures within the parasite, including hemoglobin containing structures. Treatment of the parasites with Dynasore, an inhibitor of dynamin, resulted in long hemoglobin filled tubes, suggesting that dynamin plays a role on the pinching off hemoglobin filled structures, that emanates from the end of the cytostomal tube. The authors also incubated the parasites with N-ethylmaleimide (NEM) to evaluate a possible role of SNAREs on the fusion of the hemoglobin containing structures to the food vacuole. Because of the broad effects of NEM, it is difficult to draw any definite conclusion about SNARE involvement. Nevertheless, it was reported that NEM treatment altered hemoglobin uptake pathways, resulting, as in the in the Dynasore treatment, on the formation of large hemoglobin filled tubes.

Phosphoinositides are important for regulating signaling and trafficking events in most eukaryotic cells. These phospholipids are generated as a result of phosphorylation of phosphatidylinositol by specific phosphatidylinositol (PI) kinases. In *P. falciparum*, a single PfPI3K is present and catalyzes the formation of several types of phosphoinositides (Vaid, Ranjan, Smythe, Hoppe, & Sharma, 2010). Immunofluorescence assays showed that this enzyme is localized in vesicular compartments near the parasite membrane and in the food vacuole. Inhibition of PfPI3K affected the transport of hemoglobin-containing structures to the food vacuole.

Evidence also suggests that Rab proteins play a role in maintaining and functioning the endolysosomal system. However, it remains unclear which specific Rab proteins are directly involved. (Howe, Kelly, Jimah, Hodge, & Odoma, 2013; Kennedy et al., 2019). Eleven Rabs were already identified in *P. falciparum* (Quevillon et al., 2003). Among these, Rab5a–c and Rab7 have orthologs in other organisms that are associated to endocytosis mechanisms (Spielmann, Gras, Sabitzki, & Meissner, 2020). The use of a dominant negative parasite indicated that Rab5a is involved in hemoglobin uptake (Elliott et al., 2008). Nevertheless, further analyses using a conditionally inactivated parasite for Rab5a did not confirm the previous results obtained with dominant negative parasite (Birnbaum et al., 2017). While the Rab5 equivalent in the parasite has yet to be identified, another protein, which has been shown to work with Rab5 along the endosomal pathway in other organisms (Gengyo-Ando et al., 2007; Nielsen et al., 2000) has also been demonstrated to be directly involved in hemoglobin uptake in *P. falciparum*. Conditional inactivation of Vps45 lead to accumulation of hemoglobin filled vesicles in the parasite, indicating that this protein has a role on the transport of the hemoglobin incorporated via endocytosis to the food vacuole (Jonscher et al., 2019).

Another molecular component that has been recently associated to the hemoglobin endocytosis pathway is the Kelch13 protein. The Kelch13 gene is highly conserved across malaria species, and mutations in this gene are associated with resistance to artemisinin drugs (ART) (Ariey et al., 2014; Straimer et al., 2015). ART combination therapies are currently the gold standard for malaria treatment and control. Previous studies reported that the decrease in hemoglobin uptake at *P. falciparum* ring stage leads to ART resistance because fewer hemoglobin degradation products are available to activate the drug (Xie et al., 2016). More recently, the Kelch13 protein was identified in a compartment close to the parasite food vacuole (Birnbaum et al., 2017). Upon these findings, Birnbaum et al. set out to identify Kelch13 interaction candidates and their role on hemoglobin endocytosis. Inactivation of Kelch13 or Kelch13 interaction proteins affect hemoglobin uptake only in ring stage parasites. This observation led the authors to suggest a model in which Kelch13 and the proteins within its associated compartment regulate endocytosis levels. (Birnbaum et al., 2020). Nevertheless, Kelch13 mutation is not the only mechanism reported on ART resistance. It has been reported that some *P. falciparum* parasites exhibit enhanced survival phenotype in the ring stage but lack Pfkelch13 mutations (Mukherjee et al., 2017). Investigating whether these parasites also exhibit anomalies in hemoglobin uptake, as well as further examining the impact of Kelch13 mutations on hemoglobin uptake during the trophozoite stage, is crucial. It is fascinating that cytostomes in *Plasmodium* spp. at stages other than the ring do not depend on the Kelch13 compartment to facilitate hemoglobin endocytosis.

5. The final route: hemoglobin catabolism

The food vacuole is a cytoplasmic compartment characterized by an acidic pH and specific proteases that facilitate the catabolism of hemoglobin. Due to these characteristics, the food vacuole is considered a lysosome-related organelle (Sigala & Goldberg, 2014; Wunderlich, Rohrbach, & Dalton, 2012).

However, the absence of typical lysosomal enzymes such as acid phosphatases and glycosidases suggests that this structure is specially committed to hemoglobin degradation (Francis et al., 1997).

The acidic pH of the food vacuole is maintained by two proton pumps, which are similar to those found in plants and other eukaryotic organisms: a V-type H+-ATPase, and a H+ -pyrophosphatase (Saliba et al., 2003). Hemoglobin catabolism is coordinated by a set of acidic hydrolases, including the aspartic proteases, plasmepsins I, II, and IV, histo-aspartic protease, the cysteine proteinases falcipain-2, -2′, -3, and falcilysin (Ginsburg, 2016). The trafficking of these hydrolytic enzymes to the food vacuole is not well characterized. By using a GFP-tagged protein it was shown that pro-plasmepsin II is transported through the secretory system to the cystosomal invagination, being then carried along with the endocytosed hemoglobin. At the food vacuole pro-plasmepsin II is proteolytically processed to mature plasmepsin II (Klemba, Beatty, Gluzman, & Goldberg, 2004). In a similar fashion, it has been also reported that falcipain-2 and falcipain-3 are transported to the PPM via the endoplasmic reticulum and the secretory pathway. Once at the plasma membrane these proteins are later endocytosed within hemoglobin filled structures (Subramanian, Sijwali, & Rosenthal, 2007). The mechanism by which the proton pumps are directed to the food vacuole remains unclear, but it is likely that they are also transported to specific domains in the parasite plasma membrane (Wiser, 2024). The observation that food vacuole hydrolytic enzymes and proton pumps are transported to the PPM corroborates the ongoing hypothesis that supports the existence of specific membrane domains that are more likely to occur during cytostome formation (Abu Bakar et al., 2010).

Hemoglobin catabolism at the food vacuole occurs by the sequential action of the hydrolytic enzymes, resulting in the release of small peptides and free ferrous protoporphyrin IX (Fe(II)PPIX). There is an overlap in and between the food vacuole proteolytic family functions, although the genetic disruption of plasmepsins or falcipains prevent parasite growth (Liu, Gluzman, Drew, & Goldberg, 2005). Amino acids are transported into the cytoplasm and integrated into various metabolic pathways of the parasite, whereas Fe(II)PPIX remains in the food vacuole, being rapidly oxidized to Fe(III)PPIX (heme). When separated from its protein component, heme can induce the generation of free radicals, causing molecular and cellular damage. Malaria parasites immobilize free heme in an inert form known as hemozoin crystal, thereby preventing such deleterious effects (Egan, 2002; Sigala & Goldberg, 2014).

Hemozoin formation begins with hemoglobin degradation. Small food vacuoles with hemozoin crystals are already detected in mid- to late-ring-stage parasites (Mullick et al., 2022; Wendt et al., 2016). Hemozoin crystal structure has been extensively characterized over the years (Fig. 4) (Wendt et al., 2021), however the specific mechanisms that nucleate and propagate hemozoin crystallization are still under discussion (Sigala & Goldberg, 2014). Specific heme binding proteins, which have been shown to accelerate heme crystallization in vitro, were characterized on *P. falciparum* (Jani et al., 2008; Papalexis et al., 2001; Sullivan, Gluzman, & Goldberg, 1996). Nevertheless, these proteins do not have orthologs in other malaria species (Sullivan et al., 1996). Alternatively, it has been suggested that the food vacuole membrane and its lipid–aqueous interface might have a key role on hemozoin nucleation and crystal propagation (Hempelmann, Motta, Hughes, Ward, & Bray, 2003; Kapishnikov et al., 2012; Kapishnikov et al., 2013).

Hemozoin crystals are also named as malarial pigment. Their presence in blood samples is a hallmark of malaria infection and has been used for centuries as a diagnostic tool. In addition to its diagnostic significance, hemozoin crystals represent an important drug target on malaria therapy. So far, the most successful drugs used on malaria treatment target the mechanisms related to hemoglobin catabolism and hemozoin crystal formation (Sigala & Goldberg, 2014).

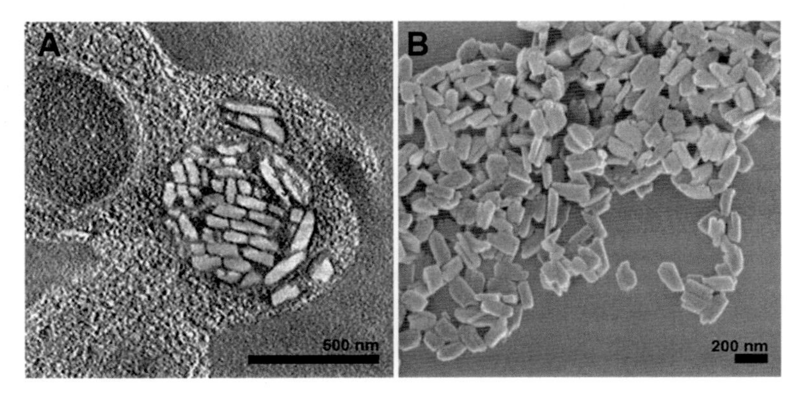

Fig. 4 Electron microscopy analysis of *P. chabaudi* hemozoin crystals. (A) Dispersion of hemozoin crystals within the food vacuole of a schizont stage parasite. Crystals are observed tightly packed inside the food vacuole. (B) Scanning electron microscopy of isolated crystals showing its characteristic brick shape observed in murine and human malaria species.

The food vacuole is not inherited by merozoites. Instead, at each cycle of infection, a new vacuole is formed (Wunderlich et al., 2012). The biogenesis, morphology, size, and number of food vacuoles is still a controversy in the literature. While some authors defend the existence of a single food vacuole (Elliott et al., 2008; Kapishnikov et al., 2012; Lazarus et al., 2008; Milani et al., 2015), others argue that the catabolism of hemoglobin can be initiated in hemoglobin filled vacuoles that are still being directed to the food vacuole (Abu Bakar et al., 2010; Hempelmann et al., 2003; Hempelmann, 2007). In fact, small vacuoles, filled with hemozoin crystals, were already observed spread throughout the cytoplasm of *P. chabaudi* (Slomianny et al., 1985; Wendt et al., 2016) and *P. falciparum* (Abu Bakar et al., 2010; Slomianny, 1990). Nevertheless, a single food vacuole is observed at the schizont stage, suggesting a yet little explored mechanism of fusion of these small food vacuoles along the parasite development. At the end of *Plasmodium* spp. intraerythrocytic development the hemozoin filled food vacuole is segregated into a residual body that is released on the blood circulation upon infected erythrocyte rupture. Released hemozoin may contribute to the inflammatory response associated with malaria (Olivier, Van Den Ham, Shio, Kassa, & Fougeray, 2014; Pham, Lamb, Deroost, Opdenakker, & Van den Steen, 2021).

6. Concluding remarks

In conclusion, the complex process of endocytosis in malaria parasites emphasizes its essential dependence on membrane trafficking. This dependency not only highlights the complexity of the parasite's survival mechanisms but also opens avenues for targeted interventions to disrupt these processes and combat malaria more effectively.

References

Abu Bakar, N., Klonis, N., Hanssen, E., Chan, C., & Tilley, L. (2010). Digestive-vacuole genesis and endocytic processes in the early intraerythrocytic stages of Plasmodium falciparum. *Journal of Cell Science, 123*, 441–450. https://doi.org/10.1242/jcs.061499.

Aikawa, M. (1971). Plasmodium: The fine structure of malarial parasites. *Experimental Parasitology, 30*, 284–320. https://doi.org/10.1016/0014-4894(71)90094-4.

Aikawa, M., Cook, R. T., Sakoda, J. J., & Sprinz, H. (1969a). Fine structure of the erythrocytic stages of Plasmodium knowlesi—A comparison between intracellular and free forms. *Zeitschrift Für Zellforschung Und Mikroskopische Anatomie, 100*, 271–284. https://doi.org/10.1007/BF00343883.

Aikawa, M., Hepler, P., Huff, C., & Sprinz, H. (1966). The feeding mechanism of avian malarial parasites. *The Journal of Cell Biology, 28*, 355–373. https://doi.org/10.1083/jcb.28.2.355.

Aikawa, M., Huff, C. G., & Sprinz, H. (1969b). Comparative fine structure study of the gametocytes of avian, reptilian, and mammalian malarial parasites. *Journal of Ultrasructure Research, 26*, 316–331. https://doi.org/10.1016/S0022-5320(69)80010-9.

Aikawa, M., Huff, C. G., & Sprinz, H. (1967). Fine structure of the asexual stages of Plasmodium elongatum. *The Journal of Cell Biology, 34*, 229–249. https://doi.org/10.1083/jcb.34.1.229.

Aikawa, M. R. C., Ito, Y., & Nijhout, M. M. (1984). New observations on gametogenesis, fertilization, and zygote transformation in Plasmodium gallinaceum. *The Journal of Protozoology, 31*, 403–413. https://doi.org/10.1111/j.1550-7408.1984.tb02987.x.

Aikawa, M., & Jordan, H. B. (1968). Fine structure of a reptilian malarial parasite. *The Journal of Parasitology, 54*, 1023. https://doi.org/10.2307/3277138.

Alcantara, C., De, L., Vidal, J. C., De Souza, W., & Cunha-e-Silva, N. L. (2014). The three-dimensional structure of the cytostome-cytopharinx complex of *Trypanosoma cruzi* epimastigotes. *Journal of Cell Science*. https://doi.org/10.1242/jcs.135491.

Ariey, F., Witkowski, B., Amaratunga, C., Beghain, J., Langlois, A.-C., Khim, N., et al. (2014). A molecular marker of artemisinin-resistant Plasmodium falciparum malaria. *Nature, 505*, 50–55. https://doi.org/10.1038/nature12876.

Bannister, L. H., Hopkins, J. M., Dluzewski, A. R., Margos, G., Williams, I. T., Blackman, M. J., et al. (2003). Plasmodium falciparum apical membrane antigen 1 (PfAMA-1) is translocated within micronemes along subpellicular microtubules during merozoite development. *Journal of Cell Science, 116*, 3825–3834. https://doi.org/10.1242/jcs.00665.

Bannister, L. H., Hopkins, J. M., Fowler, R. E., Krishna, S., & Mitchell, G. H. (2000). A brief illustrated guide to the ultrastructure of Plasmodium falciparum asexual blood stages. *Parasitology Today, 16*, 427–433. https://doi.org/10.1016/S0169-4758(00)01755-5.

Bannister, L. H., Hopkins, J. M., Margos, G., Dluzewski, A. R., & Mitchell, G. H. (2004). Three-dimensional ultrastructure of the ring stage of Plasmodium falciparum: Evidence for export pathways. *Microscopy and Microanalysis, 10*, 551–562. https://doi.org/10.1017/S1431927604040917.

Birnbaum, J., Flemming, S., Reichard, N., Soares, A. B., Mesén-Ramírez, P., Jonscher, E., et al. (2017). A genetic system to study Plasmodium falciparum protein function. *Nature Methods, 14*, 450–456. https://doi.org/10.1038/nmeth.4223.

Birnbaum, J., Scharf, S., Schmidt, S., Jonscher, E., Hoeijmakers, W. A. M., Flemming, S., et al. (2020). A Kelch13-defined endocytosis pathway mediates artemisinin resistance in malaria parasites. *Science (New York, N. Y.), 367*(1979), 51–59. https://doi.org/10.1126/science.aax4735.

Boddey, J. A., & Cowman, A. F. (2013). *Plasmodium* Nesting: Remaking the erythrocyte from the inside out. *Annual Review of Microbiology, 67*, 243–269. https://doi.org/10.1146/annurev-micro-092412-155730.

Cowman, A. F., Berry, D., & Baum, J. (2012). The cellular and molecular basis for malaria parasite invasion of the human red blood cell. *Journal of Cell Biology, 198*, 961–971. https://doi.org/10.1083/jcb.201206112.

Cowman, A. F., & Crabb, B. S. (2006). Invasion of red blood cells by malaria parasites. *Review,* 755–766. https://doi.org/10.1016/j.cell.2006.02.006.

Cowman, A. F., Healer, J., Marapana, D., & Marsh, K. (2016). Malaria: Biology and disease. *Cell, 167*, 610–624. https://doi.org/10.1016/j.cell.2016.07.055.

Dixon, M. W. A., Dearnley, M. K., Hanssen, E., Gilberger, T., & Tilley, L. (2012). Shape-shifting gametocytes: How and why does P. falciparum go banana-shaped? *Trends in Parasitology, 28*, 471–478. https://doi.org/10.1016/j.pt.2012.07.007.

Egan, T. J. (2002). Physico-chemical aspects of hemozoin (malaria pigment) structure and formation. *Journal of Inorganic Biochemistry, 91*, 19–26. https://doi.org/10.1016/S0162-0134(02)00372-0.

Elliott, D. a, McIntosh, M. T., Hosgood, H. D., Chen, S., Zhang, G., Baevova, P., et al. (2008). Four distinct pathways of hemoglobin uptake in the malaria parasite Plasmodium falciparum. *Proceedings of the National Academy of Sciences of the USA, 105*, 2463–2468. https://doi.org/10.1073/pnas.0711067105.

Elliott, J. L., Saliba, K. J., & Kirk, K. (2001). Transport of lactate and pyruvate in the intraerythrocytic malaria parasite, Plasmodium falciparum. *Biochemical Journal, 355*, 733–739. https://doi.org/10.1042/bj3550733.

Eshar, S., Dahan-Pasternak, N., Weiner, A., & Dzikowski, R. (2011). High resolution 3D perspective of Plasmodium biology: Advancing into a new era. *Trends in Parasitology, 27*, 548–554. https://doi.org/10.1016/j.pt.2011.08.002.

Francis, S. E., Sullivan, D. J., & Goldberg, D. E. (1997). Hemoglobin metabolism in the malaria parasite Plasmodium falciparum. *Annual Review of Microbiology, 51*, 97–123. https://doi.org/10.1146/annurev.micro.51.1.97.

Frischknecht, F., & Lanzer, M. (2008). The Plasmodium falciparum Maurer's clefts in 3D. *Molecular Microbiology, 67*, 687–691. https://doi.org/10.1111/j.1365-2958.2007.06095.x.

Garten, M., Beck, J. R., Roth, R., Tenkova-Heuser, T., Heuser, J., Bleck, C. K. E., et al. (2019). Contacting domains that segregate lipid from solute transporters in malaria parasites. *BioRxiv*. 863993. https://doi.org/10.1101/863993.

Gengyo-Ando, K., Kuroyanagi, H., Kobayashi, T., Murate, M., Fujimoto, K., Okabe, S., et al. (2007). The SM protein VPS-45 is required for RAB-5-dependent endocytic transport in *Caenorhabditis elegans*. *EMBO Reports, 8*, 152–157. https://doi.org/10.1038/sj.embor.7400882.

Ginsburg, H. (2016). *The biochemistry of Plasmodium falciparum. Advances in Malaria Research*. Hoboken, NJ: John Wiley & Sons, Inc. 219–290. https://doi.org/10.1002/9781118493816.ch9.

Goldberg, D. E. (1990). Hemoglobin degradation in the malaria parasite Plasmodium falciparum: An ordered pathway in a unique organelle. *Proceedings of the National Academy of Sciences of the USA, 87*, 2931–2935. https://doi.org/10.1073/pnas.87.8.2931.

Goldberg, D. E., & Zimmerberg, J. (2020). Hardly vacuous: The parasitophorous vacuolar membrane of malaria parasites. *Trends in Parasitology, 36*, 138–146. https://doi.org/10.1016/j.pt.2019.11.006.

Hanssen, E., Knoechel, C., Dearnley, M., Dixon, M. W. A., Le Gros, M., Larabell, C., et al. (2012). Soft X-ray microscopy analysis of cell volume and hemoglobin content in erythrocytes infected with asexual and sexual stages of Plasmodium falciparum. *Journal of Structural Biology, 177*, 224–232. https://doi.org/10.1016/j.jsb.2011.09.003.

Hanssen, E., Knoechel, C., Klonis, N., Abu Bakar, N., Deed, S., LeGros, M., et al. (2011). Cryo transmission X-ray imaging of the malaria parasite, P. falciparum. *Journal of Structural Biology, 173*, 161–168. https://doi.org/10.1016/j.jsb.2010.08.013.

Helms, J. B., & Zurzolo, C. (2004). Lipids as targeting signals: Lipid rafts and intracellular trafficking. *Traffic (Copenhagen, Denmark), 5*, 247–254. https://doi.org/10.1111/j.1600-0854.2004.0181.x.

Hempelmann, E. (2007). Hemozoin biocrystallization in Plasmodium falciparum and the antimalarial activity of crystallization inhibitors. *Parasitology Research, 100*, 671–676. https://doi.org/10.1007/s00436-006-0313-x.

Hempelmann, E., Motta, C., Hughes, R., Ward, S. A., & Bray, P. G. (2003). Plasmodium falciparum: Sacrificing membrane to grow crystals? *Trends in Parasitology, 19*, 23–26. https://doi.org/10.1016/S1471-4922(02)00011-9.

Howe, R., Kelly, M., Jimah, J., Hodge, D., & Odoma, A. R. (2013). Isoprenoid biosynthesis inhibition disrupts Rab5 localization and food vacuolar integrity in Plasmodium falciparum. *Eukaryotic Cell, 12*, 215–223. https://doi.org/10.1128/EC.00073-12.

Hurbain, I., & Sachse, M. (2011). The future is cold: Cryo-preparation methods for transmission electron microscopy of cells. *Biology of the Cell/Under the Auspices of the European Cell Biology Organization, 103*, 405–420. https://doi.org/10.1042/BC20110015.

Jani, D., Nagarkatti, R., Beatty, W., Angel, R., Slebodnick, C., Andersen, J., et al. (2008). HDP—A novel heme detoxification protein from the malaria parasite. *PLoS Pathogens, 4*, e1000053. https://doi.org/10.1371/journal.ppat.1000053.

Jonscher, E., Flemming, S., Schmitt, M., Sabitzki, R., Reichard, N., Birnbaum, J., et al. (2019). PfVPS45 Is required for host cell cytosol uptake by malaria blood stage parasites. *Cell Host & Microbe, 25*, 166–173.e5. https://doi.org/10.1016/j.chom.2018.11.010.

Josling, G. A., & Llinás, M. (2015). Sexual development in Plasmodium parasites: Knowing when it's time to commit. *Nature Reviews. Microbiology, 13*, 573–587. https://doi.org/10.1038/nrmicro3519.

Kapishnikov, S., Weiner, A., Shimoni, E., Guttmann, P., Schneider, G., Dahan-Pasternak, N., et al. (2012). Oriented nucleation of hemozoin at the digestive vacuole membrane in Plasmodium falciparum. *Proceedings of the National Academy of Sciences of the USA, 109*, 11188–11193. https://doi.org/10.1073/pnas.1118120109.

Kapishnikov, S., Weiner, A., Shimoni, E., Schneider, G., Elbaum, M., & Leiserowitz, L. (2013). Digestive vacuole membrane in Plasmodium falciparum infected erythrocytes: relevance to templated nucleation of hemozoin. *Langmuir: The ACS Journal of Surfaces and Colloids, 29*, 14595–14602. https://doi.org/10.1021/la402545c.

Kass, L., Willerson, D., Rieckmann, K. H., Carson, P. E., & Becker, R. P. (1971). Plasmodium falciparum gametocytes. Electron microscopic observations on material obtained by a new method. *The American Journal of Tropical Medicine and Hygiene, 20*, 187–194. https://doi.org/10.4269/ajtmh.1971.20.187.

Kennedy, K., Cobbold, S. A., Hanssen, E., Birnbaum, J., Spillman, N. J., McHugh, E., et al. (2019). Delayed death in the malaria parasite Plasmodium falciparum is caused by disruption of prenylation-dependent intracellular trafficking. *PLoS Biology, 17*, e3000376. https://doi.org/10.1371/journal.pbio.3000376.

Klemba, M., Beatty, W., Gluzman, I., & Goldberg, D. E. (2004). Trafficking of plasmepsin II to the food vacuole of the malaria parasite Plasmodium falciparum. *Journal of Cell Biology, 164*, 47–56. https://doi.org/10.1083/jcb200307147.

Krugliak, M., Zhang, J., & Ginsburg, H. (2002). Intraerythrocytic Plasmodium falciparum utilizes only a fraction of the amino acids derived from the digestion of host cell cytosol for the biosynthesis of its proteins. *Molecular and Biochemical Parasitology, 119*, 249–256. https://doi.org/10.1016/S0166-6851(01)00427-3.

Ladda, R. L. (1969). New insights into the fine structure of rodent malarial parasites. *Military Medicine, 134*, 825–865. https://doi.org/10.1093/milmed/134.9.825.

Langreth, S. G., Jensen, J. B., Reese, R. T., & Trager, W. I. (1978). Fine structure of human malaria in vitro. *Journal of Protozoology, 25*, 443–452. https://doi.org/10.1111/j.1550-7408.1978.tb04167.x.

Lauer, S., VanWye, J., Harrison, T., McManus, H., Samuel, B. U., Hiller, N. L., et al. (2000). Vacuolar uptake of host components, and a role for cholesterol and sphingomyelin in malarial infection. *The EMBO Journal, 19*, 3556–3564. https://doi.org/10.1093/emboj/19.14.3556.

Lazarus, M. D., Schneider, T. G., & Taraschi, T. F. (2008). A new model for hemoglobin ingestion and transport by the human malaria parasite Plasmodium falciparum. *Journal of Cell Science, 121*, 1937–1949. https://doi.org/10.1242/jcs.023150.

Lew, V. L., Tiffert, T., & Ginsburg, H. (2003). Excess hemoglobin digestion and the osmotic stability of Plasmodium falciparum infected red blood cells. *Blood, 101*, 4189–4194. https://doi.org/10.1182/blood-2002-08-2654.

Liu, B., Blanch, A. J., Namvar, A., Carmo, O., Tiash, S., Andrew, D., et al. (2019). Multimodal analysis of Plasmodium knowlesi-infected erythrocytes reveals large invaginations, swelling of the host cell, and rheological defects. *Cellular Microbiology, 21*, 1–16. https://doi.org/10.1111/cmi.13005.

Liu, J., Gluzman, I. Y., Drew, M. E., & Goldberg, D. E. (2005). The role of Plasmodium falciparum food vacuole plasmepsins. *Journal of Biological Chemistry, 280*, 1432–1437. https://doi.org/10.1074/jbc.M409740200.

Liu, J., Istvan, E. S., Gluzman, I. Y., Gross, J., & Goldberg, D. E. (2006). Plasmodium falciparum ensures its amino acid supply with multiple acquisition pathways and redundant proteolytic enzyme systems. *Proceedings of the National Academy of Sciences of the USA, 103*, 8840–8845. https://doi.org/10.1073/pnas.0601876103.

Martin, R. E. (2020). The transportome of the malaria parasite. *Biological Reviews, 95*, 305–332. https://doi.org/10.1111/brv.12565.

Martin, R. E., & Kirk, K. (2007). Transport of the essential nutrient isoleucine in human erythrocytes infected with the malaria parasite Plasmodium falciparum. *Blood, 109*, 2217–2224. https://doi.org/10.1182/blood-2005-11-026963.

Mauritz, J. M. A. A., Esposito, A., Ginsburg, H., Kaminski, C. F., Tiffert, T., & Lew, V. L. (2009). The homeostasis of Plasmodium falciparum-infected red blood cells. *PLoS Computational Biology, 5*(4), 11. https://doi.org/10.1371/journal.pcbi.1000339.

Mauritz, J. M. A., Seear, R., Esposito, A., Kaminski, C. F., Skepper, J. N., Warley, A., et al. (2011). X-ray microanalysis investigation of the changes in Na, K, and hemoglobin concentration in plasmodium falciparum-infected red blood cells. *Biophysical Journal, 100*, 1438–1445. https://doi.org/10.1016/j.bpj.2011.02.007.

Ménard, R., Tavares, J., Cockburn, I., Markus, M., Zavala, F., & Amino, R. (2013). Looking under the skin: The first steps in malarial infection and immunity. *Nature Reviews. Microbiology, 11*, 701–712. https://doi.org/10.1038/nrmicro3111.

Milani, K. J., Schneider, T. G., & Taraschi, T. F. (2015). Defining the morphology and mechanism of the hemoglobin transport pathway in Plasmodium falciparum-infected erythrocytes. *Eukaryotic Cell, 14*, 415–426. https://doi.org/10.1128/EC.00267-14.

Miranda, K., Girard-dias, W., Attias, M., & Souza, W. D. E. (2015). Three dimensional reconstruction by electron microscopy in the life sciences: An introduction for cell and tissue biologists. *Molecular Reproduction and Development, 882*, 530–547. https://doi.org/10.1002/mrd.22455.

Miranda, K., Wendt, C., Gomes, F., & de Souza, W. (2022). Plasmodium: Vertebrate host. Lifecycles of pathogenic protists in humans. *Microbiology Monographs, vol. 35*, 199–281. https://doi.org/10.1007/978-3-030-80682-8_5.

Mukherjee, A., Bopp, S., Magistrado, P., Wong, W., Daniels, R., Demas, A., et al. (2017). Artemisinin resistance without pfkelch13 mutations in Plasmodium falciparum isolates from Cambodia. *Malaria Journal, 16*, 195. https://doi.org/10.1186/s12936-017-1845-5.

Mullick, D., Rechav, K., Leiserowitz, L., Regev-Rudzki, N., Dzikowski, R., & Elbaum, M. (2022). Diffraction contrast in cryo-scanning transmission electron tomography reveals the boundary of hemozoin crystals in situ. *Faraday Discussions, 240*, 127–141. https://doi.org/10.1039/D2FD00088A.

Murphy, S. C., Luisa Hiller, N., Harrison, T., Lomasney, J. W., Mohandas, N., & Haldar, K. (2006). Lipid rafts and malaria parasite infection of erythrocytes. *Molecular Membrane Biology, 23*, 81–88. https://doi.org/10.1080/09687860500473440.

Murphy, S. C., Samuel, B. U., Harrison, T., Speicher, K. D., Speicher, D. W., Reid, M. E., et al. (2004). Erythrocyte detergent-resistant membrane proteins: Their characterization and selective uptake during malarial infection. *Blood, 103*, 1920–1928. https://doi.org/10.1182/blood-2003-09-3165.

Ngotho, P., Soares, A. B., Hentzschel, F., Achcar, F., Bertuccini, L., & Marti, M. (2019). Revisiting gametocyte biology in malaria parasites. *FEMS Microbiology Reviews, 43*, 401–414. https://doi.org/10.1093/femsre/fuz010.

Nielsen, E., Christoforidis, S., Uttenweiler-Joseph, S., Miaczynska, M., Dewitte, F., Wilm, M., et al. (2000). Rabenosyn-5, a novel Rab5 effector, is complexed with Hvps45 and recruited to endosomes through a fyve finger domain. *The Journal of Cell Biology, 151*, 601–612. https://doi.org/10.1083/jcb.151.3.601.

De Niz, M., & Heussler, V. T. (2018). Rodent malaria models: Insights into human disease and parasite biology. *Current Opinion in Microbiology, 46*, 93–101. https://doi.org/10.1016/j.mib.2018.09.003.

Olivier, M., Van Den Ham, K., Shio, M. T., Kassa, F. A., & Fougeray, S. (2014). Malarial pigment hemozoin and the innate inflammatory response. *Frontiers in Immunology, 5*, 1–10. https://doi.org/10.3389/fimmu.2014.00025.

Olshina, M. A., Wong, W., & Baum, J. (2012). Holding back the microfilament—Structural insights into actin and the actin-monomer-binding proteins of apicomplexan parasites. *IUBMB Life, 64*, 370–377. https://doi.org/10.1002/iub.1014.

Papalexis, Siomos, V., Campanale, M. A., Guo, N., guo, X., Kocak, G., Foley, M., et al. (2001). Histidine-rich protein 2 of the malaria parasite, Plasmodium falciparum, is involved in detoxification of the by-products of haemoglobin degradation. *Molecular and Biochemical Parasitology, 115*, 77–86. https://doi.org/10.1016/S0166-6851(01)00271-7.

Pham, T. T., Lamb, T. J., Deroost, K., Opdenakker, G., & Van den Steen, P. E. (2021). Hemozoin in malarial complications: More questions than answers. *Trends in Parasitology, 37*, 226–239. https://doi.org/10.1016/j.pt.2020.09.016.

Ponka, P. (1999). Cell biology of heme. *The American Journal of the Medical Sciences, 318*, 241–256.

Ponnudurai, T., Lensen, A. H. W., Meis, J. F. G. M., & Meuwissen, J. H. E. (1986). Synchronization of Plasmodium falciparum gametocytes using an automated suspension culture system. *Parasitology, 93*, 263–274. https://doi.org/10.1017/S003118200005143X.

Pospich, S., Kumpula, E.-P., von der Ecken, J., Vahokoski, J., Kursula, I., & Raunser, S. (2017). Near-atomic structure of jasplakinolide-stabilized malaria parasite F-actin reveals the structural basis of filament instability. *Proceedings of the National Academy of Sciences, 114*, 10636–10641. https://doi.org/10.1073/pnas.1707506114.

Preston, T. M. (1969). The Form and Function of the Cytostome-Cytopharynx of the Culture Forms of the Elasmobranch Haemoflagellate Trypanosoma raiae Laveran & Mesnil. *The Journal of Protozoology, 16*, 320–333. https://doi.org/10.1111/j.1550-7408.1969.tb02278.x.

Prudêncio, M., Rodriguez, A., & Mota, M. M. (2006). The silent path to thousands of merozoites: The Plasmodium liver stage. *Nature Reviews. Microbiology, 4*, 849–856. https://doi.org/10.1038/nrmicro1529.

Quevillon, E., Spielmann, T., Brahimi, K., Chattopadhyay, D., Yeramian, E., & Langsley, G. (2003). The Plasmodium falciparum family of Rab GTPases. *Gene, 306*, 13–25. https://doi.org/10.1016/S0378-1119(03)00381-0.

Rudzinska, M. a, & Trager, W. (1959). Phagotrophy and two new structures in the malaria parasite Plasmodium berghei. *The Journal of Biophysical and Biochemical Cytology, 6*, 103–112.

Rudzinska, M. A., & Trager, W. (1957). Intracellular phagotrophy by malaria parasites: an electron microscope study of Plasm odium lophurae. *The Journal of Protozoology, 4*, 190–199.

Saliba, K. J., Allen, R. J. W., Zissis, S., Bray, P. G., Ward, S. A., & Kirk, K. (2003). Acidification of the malaria parasite's digestive vacuole by a H^+-ATPase and a H^+-pyrophosphatase. *Journal of Biological Chemistry, 278*, 5605–5612. https://doi.org/10.1074/jbc.M208648200.

Saliba, K. J., Horner, H. A., & Kirk, K. (1998). Transport and metabolism of the essential vitamin pantothenic acid in human erythrocytes infected with the malaria parasite Plasmodium falciparum. *Journal of Biological Chemistry, 273*, 10190–10195. https://doi.org/10.1074/jbc.273.17.10190.

Scalzi, H. A., & Bahr, G. F. (1968). An electron microscopic examination of erythrocytic stages of two rodent malarial parasites, Plasmodium chabaudi and Plasmodium vinckei. *Journal of Ultrasructure Research, 24*, 116–133. https://doi.org/10.1016/S0022-5320(68)80021-8.

Sherling, E. S., & Van Ooij, C. (2016). Host cell remodeling by pathogens: The exo-membrane system in Plasmodium-infected erythrocytesa. *FEMS Microbiology Reviews, 40*, 701–721. https://doi.org/10.1093/femsre/fuw016.

Sigala, P. A., & Goldberg, D. E. (2014). The peculiarities and paradoxes of Plasmodium heme metabolism. *Annual Review of Microbiology, 68*, 259–278. https://doi.org/10.1146/annurev-micro-091313-103537.

Silvie, O., Mota, M. M., Matuschewski, K., & Prudêncio, M. (2008). Interactions of the malaria parasite and its mammalian host. *Current Opinion in Microbiology, 11*, 352–359. https://doi.org/10.1016/j.mib.2008.06.005.

Sinnis, P., & Zavala, F. (2012). The skin: Where malaria infection and the host immune response begin. *Seminars in Immunopathology, 34*, 787–792. https://doi.org/10.1007/s00281-012-0345-5.

Sinnis, P., & Zavala, F. (2008). The skin stage of malaria infection: Biology and relevance to the malaria vaccine effort. *Future Microbiology, 3*, 275–278. https://doi.org/10.2217/17460913.3.3.275.

Slomianny, C. (1990). Three-dimensional reconstruction of the feeding process of the malaria parasite. *Blood Cells, 16*, 369–378.

Slomianny, C., Prensier, G., & Charet, P. (1985). Ingestion of erythrocytic stroma by Plasmodium chabaudi trophozoites: Ultrastructural study by serial sectioning and 3-dimensional reconstruction. *Parasitology, 90*, 579–587. https://doi.org/10.1017/S0031182000055578.

Smythe, W. A., Joiner, K. A., & Hoppe, H. C. (2008). Actin is required for endocytic trafficking in the malaria parasite Plasmodium falciparum. *Cellular Microbiology, 10*, 452–464. https://doi.org/10.1111/j.1462-5822.2007.01058.x.

Soares-Medeiros, L. C., de Souza, W., Jiao, C., Barrabin, H., & Miranda, K. (2012). Visualizing the 3D architecture of multiple erythrocytes infected with Plasmodium at nanoscale by focused ion beam-scanning electron microscopy. *PLoS One, 7*, 1–9. https://doi.org/10.1371/journal.pone.0033445.

Spielmann, T., Gras, S., Sabitzki, R., & Meissner, M. (2020). Endocytosis in plasmodium and toxoplasma parasites. *Trends in Parasitology, 36*, 520–532. https://doi.org/10.1016/j.pt.2020.03.010.

Staines, H. M., Ellory, J. C., & Kirk, K. (2001). Perturbation of the pump-leak balance for Na+ and K+ in malaria-infected erythrocytes. *American Journal of Physiology. Cell Physiology, 280*. https://doi.org/10.1152/ajpcell.2001.280.6.c1576.

Steinert, M., & Novikoff, A. B. (1960). The existence of a cytostome and the occurrence of pinocytosis in the trypanosome, trypanosoma mega. *Journal of Cell Biology, 8*, 563–569. https://doi.org/10.1083/jcb.8.2.563.

Straimer, J., Gnädig, N. F., Witkowski, B., Amaratunga, C., Duru, V., Ramadani, A. P., et al. (2015). K13-propeller mutations confer artemisinin resistance in *Plasmodium falciparum* clinical isolates. *Science (New York, N. Y.), 347*(1979), 428–431. https://doi.org/10.1126/science.1260867.

Subramanian, S., Sijwali, P. S., & Rosenthal, P. J. (2007). Falcipain cysteine proteases require bipartite motifs for trafficking to the Plasmodium falciparum food vacuole. *Journal of Biological Chemistry, 282*, 24961–24969. https://doi.org/10.1074/jbc.M703316200.

Sullivan, D. J., Gluzman, I. Y., & Goldberg, D. E. (1996). Plasmodium hemozoin formation mediated by histidine-rich proteins. *Science (New York, N. Y.), 271*(1979), 219–222. https://doi.org/10.1126/science.271.5246.219.

Vaid, A., Ranjan, R., Smythe, W. A., Hoppe, H. C., & Sharma, P. (2010). PfPI3K, a phosphatidylinositol-3 kinase from Plasmodium falciparum, is exported to the host erythrocyte and is involved in hemoglobin trafficking. *Blood, 115*, 2500–2507. https://doi.org/10.1182/blood-2009-08-238972.

Ward, G. E., Miller, L. H., & Dvorak, J. A. (1993). The origin of parasitophorous vacuole membrane lipids in malaria-infected erythrocytes. *Journal of Cell Science, 106*, 237–248. https://doi.org/10.1242/jcs.106.1.237.

Wendt, C., Rachid, R., Souza, W., De, & Miranda, K. (2016). Electron tomography characterization of hemoglobin uptake in Plasmodium chabaudi reveals a stage-dependent mechanism for food vacuole morphogenesis. *Journal of Structural Biology, 194*, 171–179. https://doi.org/10.1016/j.jsb.2016.02.014.

Wendt, C., Souza, W. D. E., Pinheiro, A., Silva, L., Pinheiro, A. A., De, S., Gauvin, R., et al. (2021). High-resolution electron microscopy analysis of malaria hemozoin crystals reveals new aspects of crystal growth and elemental composition. *Crystal Growth & Design*. https://doi.org/10.1021/ACS.CGD.1C00087.

WHO. (2023). *World malaria report*.

Wiser, M. F. (2024). The digestive vacuole of the malaria parasite: A specialized lysosome. *Pathogens, 13*, 182. https://doi.org/10.3390/pathogens13030182.

Wunderlich, J., Rohrbach, P., & Dalton, J. (2012). The malaria digestive vacuole. *Frontiers in Bioscience, 4*, 1424–1448.

Xie, S. C., Dogovski, C., Hanssen, E., Chiu, F., Yang, T., Crespo, M. P., et al. (2016). Haemoglobin degradation underpins the sensitivity of early ring stage Plasmodium falciparum to artemisinins. *Journal of Cell Science, 129*, 406–416. https://doi.org/10.1242/jcs.178830.

CHAPTER THREE

How has the evolution of our understanding of the compartmentalization of sphingolipid biosynthesis over the past 30 years altered our view of the evolution of the pathway?

Assaf Biran, Tamir Dingjan, and Anthony H. Futerman*
Department of Biomolecular Sciences, Weizmann Institute of Science, Rehovot, Israel
*Corresponding author. e-mail address: tony.futerman@weizmann.ac.il

Contents

Abstract

Sphingolipids are unique among cellular lipids inasmuch as their biosynthesis is compartmentalized between the endoplasmic reticulum (ER) and the Golgi apparatus. This compartmentalization was first recognized about thirty years ago, and the current review not only updates studies on the compartmentalization of sphingolipid biosynthesis, but also discusses the ramifications of this feature for our understanding of how the pathway could have evolved. Thus, we augment some of our recent studies by inclusion of two further molecular pathways that need to be considered when analyzing the evolutionary requirements for generation of sphingolipids, namely contact sites between the ER and the Golgi apparatus, and the mechanism(s)

Current Topics in Membranes, Volume 93
ISSN 1063-5823, https://doi.org/10.1016/bs.ctm.2024.06.001

of vesicular transport between these two organelles. Along with evolution of the individual enzymes of the pathway, their subcellular localization, and the supply of essential metabolites via the anteome, it becomes apparent that current models to describe evolution of the sphingolipid biosynthetic pathway may need substantial refinement.

Abbreviations

CerS dihydroceramide synthase
CERT ceramide transport protein
CoA coenzyme A
DAG diacylglycerol
DES1 sphingolipid delta(4)-desaturase
ER endoplasmic reticulum
ERES ER exit sites
FFAT two phenylalanines in an acidic tract
GCS ceramide glucosyltransferase
GlcCer glucosylceramide
GSLs glycosphingolipids
3-KDSR 3-ketodihydrosphinganine reductase
ORP oxysterol-binding related protein
OSBP1 oxysterol-binding protein 1
PH pleckstrin homology
PI phosphatidylinositol
PI4KIIIß phosphatidylinositol 4-kinase ß
PI4P phosphatidylinositol-4-monophosphate
PKD protein kinase D
PS phosphatidylserine
SL sphingolipid
SM sphingomyelin
SMS1 sphingomyelin synthase 1
SPT serine palmitoyl transferase
TGN *trans*-Golgi network
VAPA/B vesicle-associated membrane protein-associated protein A/B

1. Introduction

Interest in sphingolipids (SLs) was re-energized in the mid-1980s by discoveries that SLs are involved in cellular signaling pathways, in an analogous fashion to the roles that phosphatidylinositol (PI) turnover was known to play in other well-established lipid signaling pathways (Michell, 1992). Thus, Yusuf Hannun showed that sphingosine inhibited protein kinase C (Hannun, Loomis, Merrill, & Bell, 1986), Richard Kolesnick showed that sphingomyelin (SM) could be hydrolyzed to ceramide via the action of sphingomyelinase (Kolesnick, 1987), and Sarah Spiegel showed that sphingosine 1-phosphate (S1P) could act as a second messenger (Zhang et al., 1991) [it was later shown that S1P can also act as a first messenger (Maceyka & Spiegel, 2014)]. Around the same time, interest in the cell biology of SLs was reignited by work from the laboratory of Richard Pagano, who showed that fluorescently-labeled, short-acyl chain derivatives of SLs appeared to be metabolized in both the endoplasmic reticulum (ER) and in the Golgi apparatus (Lipsky & Pagano, 1983, 1985a, 1985b), rather than in the ER alone, which is the case for most other cellular lipids (Pagano & Sleight, 1985). Together, and along with the roles that SLs play in inherited metabolic disorders such as lysosomal storage diseases (Futerman & Meer, 2004), these research findings provided much of the impetus for many of the studies performed over the past 30 or so years.

This review will focus on the biosynthesis of SLs within the ER and Golgi apparatus. The underlying motivation for discussing this issue now is that the notion that SL metabolism is compartmentalized between the ER and the Golgi apparatus was first reviewed 30 years ago, in 1994 (Futerman, 1994), after the first definitive data was generated showing that the initial steps in SL biosynthesis, namely the generation of ceramide, occur in the ER (Hirschberg, Rodger, & Futerman, 1993; Mandon, Ehses, Rother, Echten, & Sandhoff, 1992), whereas subsequent steps, such as the synthesis of glucosylceramide (GlcCer) (Futerman & Pagano, 1991) and the synthesis of SM (Futerman, Stieger, Hubbard, & Pagano, 1990) occur in the Golgi apparatus. Based on these three findings, a somewhat simplistic scheme (at least by today's understanding) of the compartmentalization of SL synthesis was proposed (Futerman, 1994; Hirschberg, Rodger, & Futerman, 1993) (Fig. 1). While the general details of this scheme have largely survived the passage of time, advances in the field have led to a number of modifications along with a sizable increase in our understanding of the complexity of this pathway. By way of example, it is now known that the transport of SLs both between the ER and the Golgi apparatus, and within Golgi cisternae, uses a combination of vesicular and non-vesicular pathways, and that multiple regulatory mechanisms exist which permit cells to keep a tight rein on the levels and types of SLs which are synthesized by different cells. Moreover, recent work on ER-Golgi contact sites [i.e. the molecular machines that bridge membranes (Mesmin, Kovacs, & D'Angelo, 2019)] added more details, and revealed more complexity in an already highly intricate system.

We will first give a brief overview of the SL biosynthetic pathway, then discuss its compartmentalization, and finally discuss the implications of how

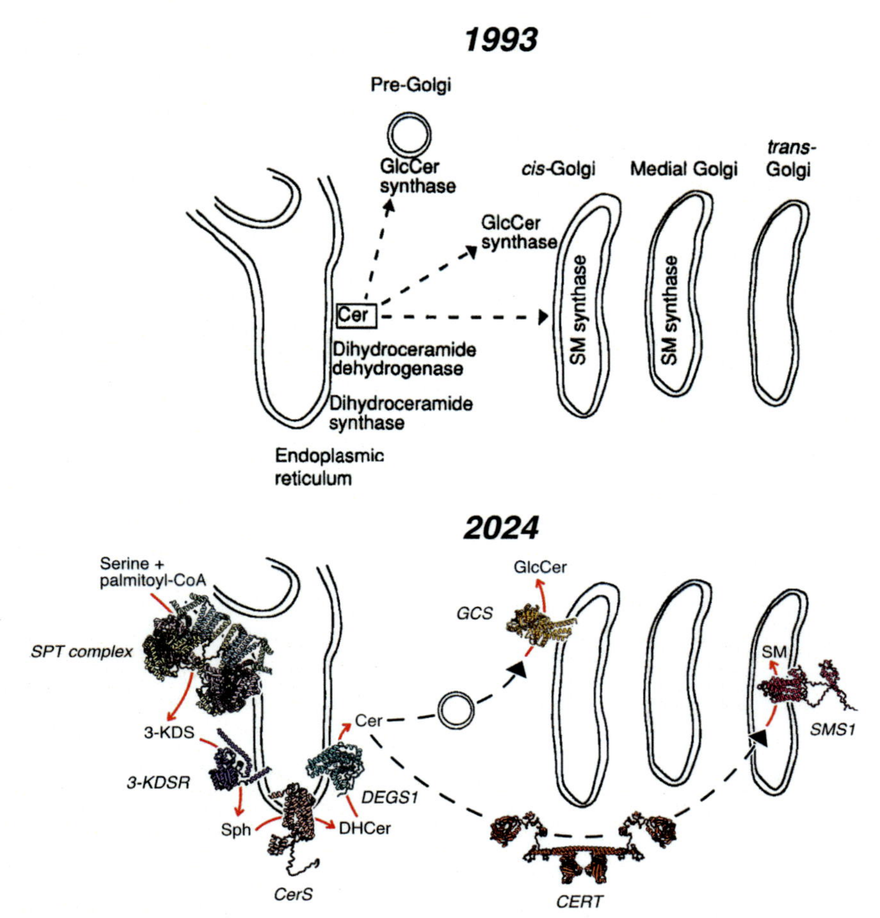

Fig. 1 Thirty years of evolution in understanding the sites and topology of the enzymes of ceramide metabolism in the ER and Golgi apparatus. The top panel shows the compartmentalization of SL biosynthesis as understood in 1994. The bottom panel shows the initial stages of the SL biosynthesis pathway as understood in 2024, with overlaid protein structure models of the SPT complex (PDB ID: 7k0m), 3-KDSR (AlphaFold model of UniProt ID: Q06136), CerS (AlphaFold model shown is CerS2; UniProt ID: Q96G23), DES1 (AlphaFold model of UniProt ID: O15121), GCS (AlphaFold model of UniProt ID: Q16739), SMS1 (AlphaFold model of UniProt ID: Q86VZ5), and CERT (model shown is from (Gehin et al., 2023); UniProt ID: Q9Y5P4). Dashed arrows indicate lipid transport routes; red arrows indicate enzyme-catalyzed reactions. Images of protein structure models generated using ChimeraX (Meng et al., 2023). *Top panel taken from Futerman, A. H. (1994). Chapter 4 Ceramide metabolism compartmentalized in theendoplasmic reticulum and Golgi apparatus. Current Topics in Membranes, 40, 93–110. https://doi.org/10.1016/s0070-2161(08)60978-8.*

progress in our understanding of the compartmentalization of SL biosynthesis over the past 30 years or so may have altered our view of the evolution of the pathway. The latter part will be, by definition, rather speculative, but discussion of how the pathway evolved is an intriguing issue for which open scientific discussion is a prerequisite to test whether the Modern synthesis (i.e. a combination of Darwinian natural selection, Mendelian inheritance, molecular genetics and population genetics) provides a sufficiently rigorous framework to make definitive conclusions about how the pathway emerged and was modified over time (Biran, Santos, Dingjan, & Futerman, 2024; Santos, Dingjan, & Futerman, 2022).

2. Overview of SL biosynthesis and its compartmentalization in the ER-Golgi complex

The first review on the compartmentalization of the SL biosynthetic pathway was published in 1994 (Futerman, 1994), with subsequent updates from our laboratory in the following years (Biran, Santos, Dingjan, & Futerman, 2024; Futerman & Riezman, 2005; Tidhar & Futerman, 2013; Zelnik, Ventura, Kim, Silva, & Futerman, 2020). Today, we know that there are four enzymes which anabolize ceramide in the ER (five if UDP-Gal:ceramide galactosyltransferase is taken into account); these four enzymes are serine palmitoyl transferase (SPT), 3-ketodihydrosphinganine reductase (3-KDSR), (dihydro)ceramide synthase (CerS) and SL delta(4)-desaturase (DES1). All subsequent steps of SL metabolism, including the formation of GlcCer via ceramide glucosyltransferase (GCS), the formation of SM via sphingomyelinase synthase 1 (SMS1), and the formation of complex glycosphingolipids (GSLs) take place in the Golgi apparatus. Later in this review, we will discuss the mechanisms by which ceramide, after its generation in the ER, is transported to the Golgi apparatus (Fig. 1).

In 1994, none of the four enzymes located in the ER had been isolated and characterized. Thus, the sites and topology of synthesis were analyzed by examination of enzymatic activity after subcellular fractionation, using tissues such as liver, which were widely used for subcellular fraction and enrichment of specific organelles. By way of example, after Pagano's work showing that C6-NBD-ceramide was apparently metabolized in the Golgi apparatus to C6-NBD-SM (Lipsky & Pagano, 1985b) it was subsequently shown that ~88% of the cellular SMS activity could be measured in an intact Golgi apparatus fraction from rat liver (Futerman, Stieger, & Hubbard, 1990).

Moreover, the lack of accessibility of SMS to proteolytic digestion suggested that the active site of SMS was located in the lumen of the Golgi apparatus. In contrast, and surprisingly at the time, the active site of GCS was accessible to proteolytic digestion in enriched Golgi fractions, suggesting that its active site faced the cytosol (Futerman & Pagano, 1991). While these two studies have generally survived the test of time, with both confirmed when SMS1 (Huitema, Dikkenberg, Brouwers, & Holthuis, 2004) and GCS (Ichikawa, Sakiyama, Suzuki, Hidari, & Hirabayashi, 1996) were eventually identified and cloned, some details were refined, such as the precise localization of SMS1 within Golgi cisternae, originally suggested to be the *cis* and *medial* cisternae (Futerman, Stieger, & Hubbard, 1990) whereas currently SM synthesis is believed to also occur in additional Golgi cisternae, including the *trans*-Golgi network (TGN) (Mizuike, Sakai, Katoh, Yamaji, & Hanada, 2023).

Some of the questions posed about the compartmentalization of SL synthesis in the ER/Golgi apparatus in 1994 have been resolved to a certain extent, whereas others remain the subject of ongoing work. One general reason suggested for this compartmentalization was that it allows "cells to regulate synthesis" by controlling the activity of the transport mechanisms required to move ceramide from the ER to the Golgi apparatus (Futerman, 1994). While the reason that the pathway of SL synthesis needs to be so tightly regulated is still a matter of debate, it must be related to the critical roles that these lipids play in cell physiology, thus demanding multiple steps of regulation (such as post-translational modifications, allosteric regulation of enzymatic activity, formation of multimeric complexes, regulation of transport and regulation of translocation), most of which have been observed for the enzymes under discussion. Moreover, advances in lipidomics analysis over the past decade or two (Ekroos, 2012) have greatly increased our appreciation of the unexpectedly large variety of lipid species in subcellular organelles, cells and tissues (Dingjan & Futerman, 2021). An open question is whether each of these SLs plays a defined role in cell physiology, or whether it is their collective behavior that is responsible for membrane function? In support of the former is the fine-tuning of membrane composition (Dingjan & Futerman, 2021) whereby each cellular membrane contains a highly selective set of lipids, which determine lipid bilayer properties (both biophysical and biochemical), and moreover, many of these lipids are asymmetrically distributed across the bilayer. The latter property has been known for some decades (den Kamp, 1979), but it is only recently that more subtle features of lipid asymmetry have been recognized, such as the asymmetric distribution of lipid acyl chain unsaturation between bilayer leaflets (Lorent et al., 2020). Lipid asymmetry can be generated not only by the action of lipid flippases or floppases (Dingjan & Futerman, 2021), but also by the topology of biosynthesis in the secretory pathway. Thus, SMS1 generates SM in the Golgi lumen, which is topologically equivalent to the outer leaflet of the plasma membrane, whereas GlcCer, after its synthesis on the cytosolic leaflet of the Golgi apparatus, is flipped by specific transporters (Biran, Santos, Dingjan, & Futerman, 2024) into the Golgi lumen. GlcCer is further metabolized before its transport to the outer leaflet of the plasma membrane, where it resides along with other SLs. Having said all of the above, a satisfactory explanation for why some steps of SL synthesis occur in the Golgi apparatus is still lacking. As discussed below, this has necessitated the emergence of highly complex mechanisms to transport ceramide from the ER to the Golgi, which presumably gives a selective advantage in evolutionary pathways (Biran, Santos, Dingjan, & Futerman, 2024), though what that advantage might be is currently completely unknown.

3. The anteome of the SL biosynthetic pathway

While the broad outline of the compartmentalization of the SL biosynthetic pathway was known 30 years ago, it is only recently that the concept of the "anteome" has been advocated. Although interconnections between metabolic pathways have been recognized for decades (as exemplified in the classic map of metabolic pathways generated by Roche [https://biochemical-pathways.com/#/map/1]), and more recently modeled by an international consortium of researchers (Thiele et al., 2013), the relationship between other metabolic pathways and the SL biosynthetic pathway has largely taken a back seat in SL research. With this in mind, we advanced the concept of the anteome, derived from the Latin prefix ante ("before") and the noun "metabolome," to describe the upstream metabolic pathways that converge upon the SL metabolic pathway. In the original study, we focused on the anteome components of the first four enzymes of the SL biosynthetic pathway (Santos, Dingjan, & Futerman, 2022). Taking into account metabolic pathways that generate amino acids (serine, glycine and alanine can all be used by SPT), pathways that generate acyl coenzyme A (CoAs), NADPH and pyridoxal-5-phosphate (PLP), which are all required by one or other of the four enzymes that generate ceramide, we concluded that SL biosynthesis requires five discrete metabolic pathways which contain up to ~10 different branches, ~28 enzymes and ~40 substrates and products. In the case of the first four enzymes, all of the anteome components come from the cytosol.

More recently, we expanded our analysis to include anteome pathways that impinge upon the metabolic steps of SL synthesis in the Golgi apparatus (Biran, Santos, Dingjan, & Futerman, 2024). Many of the components of the anteome are related to the cofactors that fuel the glycosylation steps required for GSL biosynthesis, such as activated hexoses. However, for SM synthesis, another membrane lipid is required, namely phosphatidylcholine (PC), which donates a phosphorylcholine headgroup to ceramide once ceramide is delivered to the lumen of the Golgi apparatus. Other lipids, which are discussed in more detail below, such as phosphatidylinositol phosphates (PIPs), are also required since they help target lipid transfer proteins to the right place. This analysis greatly increased the number of known pathways, enzymes, substrates and products required for SL synthesis in the ER and in the Golgi apparatus, resulting in ~10 discrete metabolic pathways, with, at minimum 55 biosynthetic enzymes and over 80 substrates and products (Biran, Santos, Dingjan, & Futerman, 2024). In addition, since many of the biosynthetic steps occur in the lumen of the Golgi apparatus, mechanisms are required to translocate either the SLs themselves, or the nucleotide sugars required for GSL synthesis, across the Golgi membrane to allow them to be accessible in the lumen. Whether these translocators and transports are de facto members of the anteome depends on how strict a definition of the anteome is applied, but irrespective of the terminology used, they are absolutely required for the synthesis of GSLs in eukaryotic cells. Thus, the misconception that describing the sequential reactions in the SL biosynthetic pathway is enough to understand how the pathways is regulated is a vast oversimplification that does not take into account the compartmentalization of SL synthesis or the role of the anteome in supplying critical metabolites, without which the pathway could not function.

4. Evolutionary implications of the complex mechanisms required to transport ceramide from the ER to the Golgi apparatus

Most studies on the evolution of the SL biosynthetic pathway have focused on phylogenetic analysis of individual enzymes in the pathway. While this is an important goal, it does not address how the pathway evolved, as a pathway, an approach which our laboratory is currently studying. However, even if we are successful in mapping the evolutionary trajectories of individual enzymes, and combining them into a comprehensive picture of how the pathway may have evolved, an additional layer of mechanistic understanding is required, which takes into account the coupling of upstream anteome pathways to the SL biosynthetic enzymes discussed earlier (Biran, Santos, Dingjan, & Futerman, 2024; Santos, Dingjan, & Futerman, 2022). Furthermore, we now add another layer of intricacy, by considering two additional components required for SL biosynthesis that we have not discussed before, namely contact sites between the ER and the Golgi apparatus and vesicular transport between these two organelles, both of which are indispensable for SL biosynthesis, at least in modern cells. The ceramide that is transported to the Golgi apparatus for SM synthesis is transported via CERT (ceramide ER transport protein), which is part of the molecular machinery that makes up ER-Golgi contact sites, whereas the ceramide that is transported for GlcCer synthesis is transported via vesicular transport. Both ER-Golgi contact sites and vesicular transport are highly complex processes, which will be described below, before considering current ideas about how these pathways evolved and how their evolution could have occurred concomitantly with the SL biosynthetic pathway and its anteome.

4.1 Vesicular transport

Vesicular transport is the process by which cargo is trafficked from one organelle to another, within the endomembrane system of eukaryotic cells, via the budding-off of vesicles from the donor organelle (Fig. 2A) followed by their transport and fusion with the target organelle. The pathway of vesicular transport has been extensively studied, starting with the pioneering work of De Duve, followed by Palade, Blobel, and subsequently by Rothman, Südhof and Schekman, each of whom received the Nobel prize for their efforts.

As might be expected, our appreciation of the complexity of the pathway has increased rapidly with time, such that the pathway of vesicular transport is now recognized to be one of the most highly regulated pathways in cell biology. Briefly, a carrier vesicle is formed by the action of a small GTPase and its activator, which recruit coat proteins that induce the formation of the vesicle (COPII for ER to Golgi transport; Fig. 2A). Once the vesicles bud-off from their membrane of origin, they move toward their target membrane either through diffusion or via binding a motor

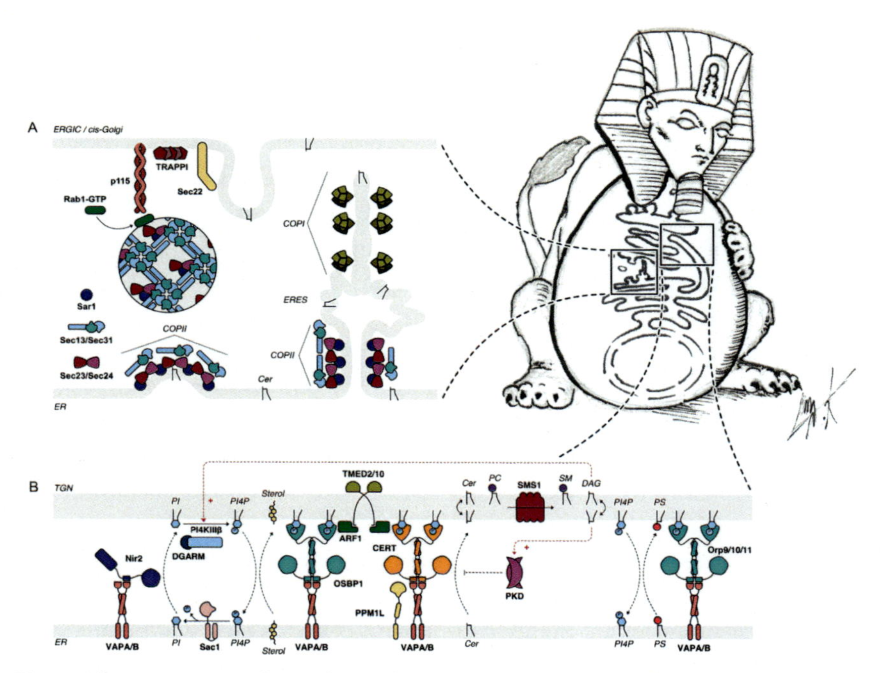

Fig. 2 The transport of SLs from the ER to the Golgi apparatus. SLs are transported between the ER and the Golgi apparatus via two highly intricate mechanisms, the first involving vesicular transport (A) and the second involving membrane contact sites (B). The *right-hand* image addresses the issue of what came first in evolutionary history, the Sphinx (SLs are named after the Greek Sphinx) or the components of the SL anteome that are absolutely required for SL biosynthesis. We now expand on the "Sphinx and the egg" conundrum by broadening the question of which came first, the compartmentalization of SL biosynthesis between the ER and the Golgi apparatus or the generation of the two independent transport mechanisms described herein. Most of components in this scheme are discussed in detail in the text but some additional details are given in the figure. Thus, multiple lipid trafficking proteins bind to both the ER and the Golgi apparatus via recognition of common targeting motifs at the ER-TGN contact site. At the ER cytoplasmic leaflet, VAPA and VAPB are involved in contact with other organelles, which is accomplished via binding proteins containing a FFAT motif. Proteins are localized to the TGN cytoplasmic leaflet via recognition of PI4P by PH domains (Godi et al., 2004). PI4P is generated at the TGN cytoplasmic leaflet by PI4KIIIß, setting up a concentration gradient for counter-transport of sterols (by OSBP1) or PS (by Orp9/10/11) from the ER. Once PI4P is deposited in the ER cytoplasmic leaflet, it is dephosphorylated by Sac1, regenerating PI in the ER. Newly-generated PI is returned to the TGN by PITP proteins, which transport PI from the ER to organellar membranes including the Golgi apparatus. The family includes proteins featuring a single lipid-binding domain (PITPa, PITPß, and PITNC1), and multi-domain membrane-associated transporters containing FFAT motifs (Nir2 and Nir3). Nir2 acts at ER-TGN contact sites. Following transport to the cytoplasmic leaflet of the TGN by CERT, ceramide flips to the lumenal leaflet where it is converted to SM by SMS1,

protein on a cytoskeletal track. At the target membrane, the vesicles are tethered by tethering factors and Rab GTPases, and subsequently fuse with the membrane via the action of SNARE proteins, thus delivering their cargo to the target organelle (Cai, Reinisch, & Ferro-Novick, 2007; Lujan & Campelo, 2021) (Fig. 2A). This pathway is relevant for SL biosynthesis since ceramide, which is generated in the ER, is transported via vesicular transport to the Golgi apparatus for its subsequent metabolism to GlcCer at the cytosolic surface, raising the question of what came first in evolutionary history, generation of the compartmentalization of SL synthesis with non-vesicular ceramide transport routes or the generation of vesicular transport between the ER and the Golgi apparatus. Interestingly, in yeast, delivery of ceramide from the ER to the Golgi is not completely dependent on vesicular transport (Funato & Riezman, 2001).

Prior to their entry into transport vesicles that bud-off from the ER, vesicle components (either integral membrane proteins or soluble cargo) are targeted via a sorting mechanism to ER exit sites (ERES), a distinct part of the ER membrane, comprising distinct lipids (Rodriguez-Gallardo et al., 2020; Tang & Ginsburg, 2023; Weigel et al., 2021). After this, vesicles are formed by the small GTPase Sar1, which recruits Sec31 and Sec13, which together form COPII, the coat protein complex (Fig. 2A) (Rout & Field, 2016; Tang & Ginsburg, 2023; Weigel et al., 2021). Once the coated vesicle is formed at the ERES, there are two models that could potentially explain how cargo is transported to the *cis*-Golgi (Tang & Ginsburg, 2023), either the formation of discrete vesicles or the formation of tubules that directly connect the ER with the Golgi apparatus. In the first, a coated vesicle buds off the ER membrane and either moves toward the ER-Golgi intermediate compartment (ERGIC), which subsequently fuses with the *cis*-Golgi membrane (Lorente-Rodríguez & Barlowe, 2011) or moves directly to the *cis*-Golgi. In either case the vesicle is tethered by TRAPPI/Rab1-GTP/p115 (Cai, Reinisch, & Ferro-Novick, 2007) and undergoes fusion via the SNARE protein Sec22 (Lujan & Campelo, 2021) (Fig. 2A). In the alternative model, i.e. the formation of continuous tubules, COPII remains tethered at the base of the ERES complex

producing DAG as a by-product. DAG translocates to the TGN cytoplasmic leaflet, where it recruits PKD which in turn down-regulates CERT activity and activates PI4KIIIß. In this way, SL transport at ER-TGN contact sites is closely involved with local PI/PI4K flux. *The right-hand image is adapted from Biran, A., Santos, T. C. B., Dingjan, T., & Futerman, A. H. (2024). The Sphinx and the egg: Evolutionary enigmas of the(glyco) sphingolipid biosynthetic pathway. Biochimica et Biophysica Acta (BBA)—Molecular and Cell Biology of Lipids, 1869, 159462. https://doi.org/10.1016/j.bbalip.2024.159462.*

(Fig. 2A), which can form an interwoven network of tubes (Weigel et al., 2021). The ERES then extends an intermediate tubule, through which the secreted vesicle components travel to the Golgi apparatus. The intermediate tubule is bound to another coat protein complex, COPI, and travels toward the Golgi apparatus through microtubule-directed movement (Weigel et al., 2021).

Irrespective of the precise details of vesicle formation, membrane lipids and in particular SLs, play a critical role in protein transport from the ER to the Golgi apparatus. For instance, the binding of coat-forming-proteins is often triggered and accompanied by deformation and curvature of membranes (Lippincott-Schwartz & Phair, 2010; Rout & Field, 2016; Zelnik et al., 2020). SLs have unique biophysical properties that affect membrane shape (Zelnik et al., 2020) such that their concentration across the membrane and at specific sites in the plane of the lipid bilayer, along with the concentration of cholesterol and GSLs, affects the ability of the membrane to bend and form vesicles (Lippincott-Schwartz & Phair, 2010). Thus, vesicle formation depends on the very lipids that need to be delivered by the vesicles. Moreover, SLs with different acyl chain lengths can directly affect ERES (Rodriguez-Gallardo et al., 2020). In yeast, the type of ERES used by transmembrane proteins contain long acyl-chain SLs (C16:0–C18:0), while glycosylphosphatidylinositol-anchored proteins use specialized ERES composed of very-long acyl chain SLs (C26:0). Moreover, the sorting of glycosylphosphatidylinositol-anchored proteins to specialized ERES depends on the very-long chain SLs. Thus, not only does the formation of vesicles depends on SLs, but the content of the vesicle also depends on the chain length of the SL.

The relationship between the evolution of the molecular machinery of vesicle formation and the generation of the SL biosynthetic pathway is unknown, and few, if any, testable or conceptual theories have been generated to examine this puzzle. Some work has been done to examine the evolution of trafficking pathways, with the last eukaryotic common ancestor suggested to contain such a system (Barlow, Nývltová, Aguilar, Tachezy, & Dacks, 2018; Rout & Field, 2016). However, even though some building blocks of the trafficking system have been found in Asgard archaea (Rout & Field, 2016) [considered the closest prokaryotic relative of eukaryotes (Imachi et al., 2020)], critical proteins, such as coat forming proteins, are not found in Asgard (Rout & Field, 2016), although they have been found in some bacteria (Ferrelli et al., 2023). Thus, we are left with two highly complex systems (SL biosynthesis and vesicular trafficking) that in modern cells show a high degree of interdependence, raising intriguing questions about whether their co-evolution was required or whether the two pathways evolved independently. For both of these, evolutionary trajectories need to be postulated (Biran, Santos, Dingjan, & Futerman, 2024) to reduce to a minimum the conjecture which often accompanies evolutionary biology. The evolutionary challenges become even more convoluted when considering the alternative mechanism of moving material from the ER to the Golgi, namely via contact sites.

4.2 ER-Golgi contact sites

Until 1999, it was assumed that all SLs are transported between the ER and the Golgi apparatus via the kind of vesicular transport mechanisms described above (Futerman, 1994). However, this changed after the discovery of the lipid transfer protein CERT by the laboratory of Kentaro Hanada, which is required for SM synthesis via SMS1 in the lumen of the Golgi apparatus (Fukasawa, Nishijima, & Hanada, 1999; Hanada et al., 2003). More recently, CERT was shown to bridge ER-Golgi contact sites via its N-terminal pleckstrin-homology (PH) domain and two phenylalanines in an acidic tract (FFAT) motif, while the C-terminal START domain of CERT transports ceramide from the ER to both the *trans*-Golgi and TGN (Fig. 2B) (Hanada et al., 2003; Kumagai & Hanada, 2019). That CERT acts at membrane contact sites, rather than by transporting single molecules of ceramide between the ER and the Golgi apparatus, is not surprising in retrospect in light of the likely inefficiency of the former and in light of the large number of contact sites detected between the ER and the Golgi apparatus.

Roughly a quarter of the surface area of the TGN is in contact with the ER (Venditti et al., 2019b), with the two organelles in very close proximity (~10–20 nm) at these sites (Fig. 2B) (Matteis & Rega, 2015). ER-Golgi contact sites, specifically in metazoa, have a variety of functions, most notably lipid exchange via proteins tethered to both organelles (David, Castro, & Schuldiner, 2021). The transfer of lipids at these tethered ER-Golgi contact sites can regulate lipid-based signaling pathways, adjust the local membrane environment, and provide lipids for the biogenesis of organellar membranes (Mesmin, Kovacs, & D'Angelo, 2019; Prinz, Toulmay, & Balla, 2020). One possible key functional aspect of ceramide trafficking via ER-TGN contact sites is that this allows ceramide to bypass glycosylation at the *cis*-Golgi. Direct trafficking of ceramide from the ER to the TGN might therefore constitute a metabolic shortcut, providing an entirely separate pool of ceramide for the production of SM. Another key function is that by bypassing the *cis*- and medial-Golgi, a bigger pool of SLs can form in the TGN, which is important for vesicular transport into and out of the TGN (Barman et al., 2022; Mesmin, Kovacs, & D'Angelo, 2019).

CERT is regulated by a number of kinases and phosphatases at ER-TGN contact sites (Fig. 2B) (Weber, Hornjik, Olayioye, Hausser, & Radde, 2015). The de novo synthesis of SM produces diacylglycerol (DAG) as a by-product in the TGN bilayer (Pagano, 1988); DAG recruits protein kinase D (PKD) (Baron & Malhotra, 2002) which inhibits CERT by phosphorylation in a region adjacent to the N-terminal PH domain, interfering with the binding of phosphatidylinositol-4-monophosphate (PI4P) (Mesmin, Kovacs, & D'Angelo, 2019). PKD localization to the TGN is reduced by down-regulating SMS1 (Villani et al., 2008), underscoring the tight integration of CERT activity with SM biosynthesis at ER-TGN contact sites. CERT activation by protein phosphatase 1L (PPM1L) also occurs at the ER-TGN contact site; PPM1L is a phosphatase that binds to vesicle-associated membrane protein-associated protein A (VAPA) at ER-Golgi contact sites and activates CERT by dephosphorylating the same serine repeat motif located between the PH domain and FFAT motif phosphorylated by PKD (Saito et al., 2008). PP1ML is recruited to ER-TGN contact sites by acyl-CoA-binding domain-containing protein 3 (Shinoda et al., 2012). Thus, these CERT-regulating enzymes should be considered as critical players in the regulation of the SL biosynthetic pathway.

A supercomplex of CERT with another well-studied lipid transport protein, oxysterol-binding protein 1 (OSBP1) has also been proposed. OSBP1 transports sterols from the ER to the TGN and transports PI4P in the reverse direction through its lipid binding domain (an OSBP-related domain or ORD) (Fig. 2B) (Mesmin et al., 2013). Much like CERT, OSBP1 bridges the ER-TGN contact site via its PH domain and FFAT motif. The supercomplex of these two lipid transport proteins formed by physical interactions with transmembrane emp24 domain-containing protein 2/10 (TMED2/10) appears to control lipid composition in the plasma membrane (Anwar et al., 2022). Here again, regulation of the supercomplex occurs at the ER-TGN contact site; the kinase phosphatidylinositol 4-kinase ß (PI4KIIIß) phosphorylates PI to produce PI4P in the Golgi apparatus, maintaining the PI4P concentration gradient necessary for OSBP1-mediated sterol transport. Degradation of PI4P by the ER-resident 4-phosphatase, Sac1 also occurs within the contact site (Venditti et al., 2019a). Regulation of PI/PI4P flux is relevant to SL biosynthesis since inhibition of PI4KIIIß prevents de novo SM synthesis (Tóth et al., 2006). PI4KIIIß was recently found to be recruited to ER-TGN contact sites by C10orf76 [renamed DGARM (Mizuike & Hanada, 2024)], and depletion of this protein in HeLa cells disrupts CERT localization at ER-TGN contact sites and impairs SM synthesis (Mizuike & Hanada, 2024). PI4KIIIß is stimulated by GTPase ADP-ribosylation factor (ARF1) (Mesmin, Kovacs, & D'Angelo, 2019), a multifunctional GTPase that also recruits OSBP1 to ER-TGN contact sites (Levine & Munro, 2002). SL biosynthesis and transport is thus integrated into the regulation of ER-TGN contact sites by PI4K flux (Posor, Jang, & Haucke, 2022) and physical association with OSBP1 (Fig. 2B) (Posor, Jang, & Haucke, 2022).

An additional role for oxysterol-binding related proteins (ORPs) has recently been suggested based on their roles in regulating SM biosynthesis (Cabukusta et al., 2023). ORPs contain an N-terminal PH domain (Olkkonen et al., 2006) and OSBP1, ORP1, 2, 3, 4, 6, 7 and 9 contain a FFAT motif (Loewen, Roy, & Levine, 2003; Mesmin et al., 2013). Multiple ORPs show affinity for sterols (Kentala, Weber-Boyvat, & Olkkonen, 2016), and a number have also been reported to transport phosphatidylserine (PS) (including ORP5 and ORP10) (Maeda et al., 2013). ORP9 is required to maintain Golgi structure, and at least one of OSBP1 or ORP9 is required to stabilize the ER-TGN contact site; the PS transport activity of ORP10 is likewise required to maintain the ER-TGN contact site (Fig. 2B) (Cai et al., 2007). A recent study suggested that an ORP9-ORP11 heterodimer localized to the ER-TGN contact site transports PS from the ER to the Golgi apparatus, and in exchange transports PI4P in the reverse direction. Knockout of this heterodimer is associated with decreased de novo synthesis of SM similar to CERT knockouts, suggesting an important role for PS-PI4P exchange at the ER-TGN contact site in SL biosynthesis (Fig. 2B).

Similar to the pathway of vesicular trafficking, little is known about how ER-TGN contact sites evolved and how this might be related to the emergence of SL biosynthesis. CERT is broadly absent in bacteria, plants, and fungi, as is SM, suggesting a co-evolutionary emergence for CERT and SM, which would, from the outset, insulate SM synthesis from competing metabolic reactions (Hanada, 2014). The mutational trajectory allowing these two proteins to emerge together is of course unknown. Moreover, how does the emergence of ER-TGN contact sites fit into the developmental history of SL biosynthesis? Some components may be able to emerge independently: ER-TGN contact sites are not destabilized by depleting CERT, but rather require multiple members of the ORP family as noted above. Intriguingly, OSBP1 orthologs are present in most eukaryotes, while CERT is limited to metazoa (D'Angelo, Vicinanza, & Matteis, 2008), although ceramide transport is not. Indeed, two CERTs have been found in *Saccharomyces cerevisiae* (Limar et al., 2023). However, neither of the two transport proteins are homologous to CERT, suggesting that ceramide transport may have occurred by convergent evolution. Why have CERTs apparently emerged multiple times? A simple answer might be that the differences between the metazoan and non-metazoan Golgi apparatus (Benvenuto et al., 2024) require a different solution for ceramide transport. However, this answer overlooks the conservation of OSBP1 and ORP proteins across eukaryota. SL metabolism seems to feature multiple cases of convergent evolution, including the emergence of biosynthetic proteins that converged between different bacteria, or between bacteria and eukaryotes (Dhakephalkar, Stukey, Guan, Carman, & Klein, 2023; Heaver et al., 2022; Stankeviciute et al., 2022). Attempts to determine why and how proteins with similar functions, but totally different sequences should have emerged multiple times, and specifically in the context of SL biosynthesis, are currently little more than speculation.

5. Concluding remarks

Generating a SL in a eukaryotic cell is not a simple proposition. Not only do the enzymes in the metabolic pathway need to be adapted to their function, but substrates and co-factors need to be supplied in such a way that they are easily accessible to the active sites of the enzymes. With this in mind, we recently introduced the notion of the anteome, which we defined as those metabolic pathways which impinge directly on the metabolic pathway being studied, in this case the SL biosynthetic pathway (Biran, Santos, Dingjan, & Futerman, 2024; Santos, Dingjan, & Futerman, 2022). In addition, and the focus of the current review, is the need to transport ceramide between the ER and the Golgi apparatus, which occurs via two independent mechanisms, with both being highly complex as illustrated in Fig. 2.

Clearly, our thinking about the SL biosynthetic pathway has undergone a process of evolution over the past 30 years or so. This raises the question of whether the evolution of our understanding of the pathway has led to advances in our understanding of how the pathway could have evolved? Rather, could it be that our appreciation of the complexity of the SL biosynthesis has actually raised more questions than can currently be answered? To quote the late John Gall, "A complex system that works is invariably found to have evolved from a simple system that works. The inverse proposition also appears to be true: A complex system designed from scratch never works and cannot be made to work" (Gall, 1977). Are we able to generate the complex systems of the SL biosynthetic pathway described above by proposing modifications of more simple pathways?

A typical explanation for how complex metabolic systems might have evolved from simpler ones would be that the simpler systems originally performed a different function to that performed once the more complex system was formed. While this suggestion appears to be conceptually attractive, it does not provide any mechanistic or systematic details as to how such events could have occurred, and specifically, does not provide any kind of pathway by which mutations could have become fixed in a stepwise fashion in the population to give a selective advantage (although the possibility that such a mutation is neutral also needs to be taken into account). Indeed, very few mutational trajectories have been defined for the evolution of any metabolic pathway, just as few details are available to suggest mutational trajectories for individual enzymes of the SL biosynthetic pathway. By way of example, in the SL biosynthetic pathway, the first enzyme, SPT, belongs to a family of α-oxoamine synthases, which are believed to have mutated into an enzyme with the capability to generate a sphingoid long chain base using an acyl CoA and an amino acid (Yard et al., 2007). Assuming that this mutational step occurred, did the generation of the sphingoid long chain base itself give a selective advantage, or did it require the co-evolution of 3-KDSR, the next enzyme in the pathway? Moving further down the pathway, did the evolution of CerS and of DES1 depend on the ability to co-evolve the machinery needed to remove ceramide from the ER and to deliver it to the Golgi apparatus, or did the two pathways described in this review evolve independently and were subsequently co-opted by the SL biosynthetic pathway? Did SMS1 evolve prior to generation of the transport mechanisms between the ER and the Golgi or did it perhaps evolve in the ER (where it has a homolog, SMSr (Vacaru et al., 2009)), and then sequester a targeting signal to allow it to be localized to the Golgi? Many other similar questions and putative scenarios can be proposed, but in order for them to be critically appraised, detailed research programs are required; a good start in such a program would be to determine the putative mutational trajectories of individual enzymes of the pathway and subsequently to determine whether these mutational trajectories could be knit together into a scheme that could predict how the pathway may have evolved as a pathway.

It might be construed that we are being obstructionist by asking such questions about how the SL biosynthetic pathway evolved; after all, we are here so we must have got here somehow, and this must have involved an evolutionary process of some kind. While the latter is absolutely true, it does not address the details of how these evolutionary events could have occurred, and as discussed above, the details in the case of the SL biosynthetic pathway are perplexing to say the least. What is more clear is that advances in our understanding of this pathway have generated many unexpected questions that could not have been anticipated beforehand, which will keep SL researchers busy for at least another 30 years, if not longer.

Funding

Work in the Futerman laboratory on SL synthesis and complexity is supported by the Walt and Rowena Shaw Foundation.

Declaration of competing interest

The authors declare that they have no known competing financial interests or personal relationships that could have appeared to influence the work reported in this paper.

Data availability

No data was used for the research described in the article.

References

Anwar, M. U., Sergeeva, O. A., Abrami, L., Mesquita, F. S., Lukonin, I., Amen, T., et al. (2022). ER-Golgi-localized proteins TMED2 and TMED10 control the formation of plasma membrane lipid nanodomains. *Developmental Cell, 57*(2334–2346), e8. https://doi.org/10.1016/j.devcel.2022.09.004.

Barlow, L. D., Nývltová, E., Aguilar, M., Tachezy, J., & Dacks, J. B. (2018). A sophisticated, differentiated Golgi in the ancestor of eukaryotes. *BMC Biology, 16*, 27. https://doi.org/10.1186/s12915-018-0492-9.

Barman, B., Sung, B. H., Krystofiak, E., Ping, J., Ramirez, M., Millis, B., et al. (2022). VAP-A and its binding partner CERT drive biogenesis of RNA-containing extracellular vesicles at ER membrane contact sites. *Developmental Cell, 57*, 974–994.e8. https://doi.org/10.1016/j.devcel.2022.03.012.

Baron, C. L., & Malhotra, V. (2002). Role of diacylglycerol in PKD recruitment to the TGN and protein transport to the plasma membrane. *Science (New York, N. Y.), 295*, 325–328. https://doi.org/10.1126/science.1066759.

Benvenuto, G., Leone, S., Astoricchio, E., Bormke, S., Jasek, S., D'Aniello, E., et al. (2024). Evolution of the ribbon-like organization of the Golgi apparatus in animal cells. *Cell Reports, 43*, 113791. https://doi.org/10.1016/j.celrep.2024.113791.

Biran, A., Santos, T. C. B., Dingjan, T., & Futerman, A. H. (2024). The Sphinx and the egg: Evolutionary enigmas of the (glyco)sphingolipid biosynthetic pathway. *Biochimica et Biophysica Acta (BBA) – Molecular and Cell Biology of Lipids, 1869*, 159462. https://doi.org/10.1016/j.bbalip.2024.159462.

Cabukusta, B., Pauwels, S. B., Akkermans, J. J. L. L., Blomberg, N., Mulder, A. A., Koning, R. I., et al. (2023). A lipid transfer protein knockout library reveals

ORP9-ORP11 dimer mediating PS/PI(4)P exchange at the ER-trans Golgi contact site to promote sphingomyelin synthesis. 2023.06.02 *bioRxiv*543249. https://doi.org/10.1101/2023.06.02.543249.

Cai, H., Reinisch, K., & Ferro-Novick, S. (2007). Coats, tethers, rabs, and SNAREs work together to mediate the intracellular destination of a transport vesicle. *Developmental Cell, 12*, 671–682. https://doi.org/10.1016/j.devcel.2007.04.005.

David, Y., Castro, I. G., & Schuldiner, M. (2021). The fast and the furious: Golgi contact sites. *Contact (Geneva, Switzerland), 4*, 25152564211034424. https://doi.org/10.1177/25152564211034424.

den Kamp, J. A. F. O. (1979). Lipid asymmetry in membranes. *Annual Review of Biochemistry, 48*, 47–71. https://doi.org/10.1146/annurev.bi.48.070179.000403.

Dhakephalkar, T., Stukey, G. J., Guan, Z., Carman, G. M., & Klein, E. A. (2023). Characterization of an evolutionarily distinct bacterial ceramide kinase from *Caulobacter crescentus*. *Journal of Biological Chemistry, 299*, 104894. https://doi.org/10.1016/j.jbc.2023.104894.

Dingjan, T., & Futerman, A. H. (2021). The fine-tuning of cell membrane lipid bilayers accentuates their compositional complexity. *BioEssays: News and Reviews in Molecular, Cellular and Developmental Biology*, e2100021. https://doi.org/10.1002/bies.202100021.

D'Angelo, G., Vicinanza, M., & Matteis, M. A. D. (2008). Lipid-transfer proteins in biosynthetic pathways. *Current Opinion in Cell Biology, 20*, 360–370. https://doi.org/10.1016/j.ceb.2008.03.013.

Ekroos, K. (2012). Lipidomics perspective: From molecular lipidomics to validated clinical diagnostics. In K. Ekroos (Ed.). *Lipidomics* (pp. 1–19)Wiley.

Ferrelli, M. L., Pidre, M. L., García-Domínguez, R., Alberca, L. N., Saz-Navarro, D. D., Santana-Molina, C., et al. (2023). Prokaryotic membrane coat – like proteins: An update. *Journal of Structural Biology, 215*, 107987. https://doi.org/10.1016/j.jsb.2023.107987.

Fukasawa, M., Nishijima, M., & Hanada, K. (1999). Genetic evidence for ATP-dependent endoplasmic reticulum-to-Golgi apparatus trafficking of ceramide for sphingomyelin synthesis in Chinese hamster ovary cells. *The Journal of Cell Biology, 144*, 673–685. https://doi.org/10.1083/jcb.144.4.673.

Funato, K., & Riezman, H. (2001). Vesicular and nonvesicular transport of ceramide from ER to the Golgi apparatus in yeast. *The Journal of Cell Biology, 155*, 949–960. https://doi.org/10.1083/jcb.200105033.

Futerman, A. H. (1994). Chapter 4 Ceramide metabolism compartmentalized in the endoplasmic reticulum and Golgi apparatus. *Current Topics in Membranes, 40*, 93–110. https://doi.org/10.1016/s0070-2161(08)60978-8.

Futerman, A. H., & Meer, G. V. (2004). The cell biology of lysosomal storage disorders. *Nature Reviews Molecular Cell Biology, 5*, 554–565. https://doi.org/10.1038/nrm1423.

Futerman, A. H., & Pagano, R. E. (1991). Determination of the intracellular sites and topology of glucosylceramide synthesis in rat liver. *Biochemical Journal, 280(Pt 2)*, 295–302.

Futerman, A. H., & Riezman, H. (2005). The ins and outs of sphingolipid synthesis. *Trends in Cell Biology, 15*, 312–318. https://doi.org/10.1016/j.tcb.2005.04.006.

Futerman, A. H., Stieger, B., Hubbard, A. L., & Pagano, R. E. (1990). Sphingomyelin synthesis in rat liver occurs predominantly at the cis and medial cisternae of the Golgi apparatus. *The Journal of Biological Chemistry, 265*, 8650–8657.

Gall, J. (1977). *Systemantics: How systems work and especially how they fail*. New York: Quadrangle/New York Times Book Co.

Gehin, C., Lone, M. A., Lee, W., Capolupo, L., Ho, S., Adeyemi, A. M., et al. (2023). CERT1 mutations perturb human development by disrupting sphingolipid homeostasis. *Journal of Clinical Investigation, 133*, e165019. https://doi.org/10.1172/jci165019.

Godi, A., Campli, A. D., Konstantakopoulos, A., Tullio, G. D., Alessi, D. R., Kular, G. S., et al. (2004). FAPPs control Golgi-to-cell-surface membrane traffic by binding to ARF and PtdIns(4)P. *Nature Cell Biology, 6*, 393–404. https://doi.org/10.1038/ncb1119.

Hanada, K. (2014). Co-evolution of sphingomyelin and the ceramide transport protein CERT. *Biochimica et Biophysica Acta (BBA) – Molecular and Cell Biology of Lipids, 1841*, 704–719. https://doi.org/10.1016/j.bbalip.2013.06.006.

Hanada, K., Kumagai, K., Yasuda, S., Miura, Y., Kawano, M., Fukasawa, M., et al. (2003). Molecular machinery for non-vesicular trafficking of ceramide. *Nature, 426*, 803–809. https://doi.org/10.1038/nature02188.

Hannun, Y. A., Loomis, C. R., Merrill, A. H. J., & Bell, R. M. (1986). Sphingosine inhibition of protein kinase C activity and of phorbol dibutyrate binding in vitro and in human platelets. *The Journal of Biological Chemistry, 261*, 12604–12609.

Heaver, S. L., Le, H. H., Tang, P., Baslé, A., Barone, C. M., Vu, D. L., et al. (2022). Characterization of inositol lipid metabolism in gut-associated Bacteroidetes. *Nature Microbiology, 7*, 986–1000. https://doi.org/10.1038/s41564-022-01152-6.

Hirschberg, K., Rodger, J., & Futerman, A. H. (1993). The long-chain sphingoid base of sphingolipids is acylated at the cytosolic surface of the endoplasmic reticulum in rat liver. *Biochemical Journal, 290(Pt 3)*, 751–757.

Huitema, K., Dikkenberg, J. V. D., Brouwers, J. F., & Holthuis, J. C. (2004). Identification of a family of animal sphingomyelin synthases. *The EMBO Journal, 23*, 33–44.

Ichikawa, S., Sakiyama, H., Suzuki, G., Hidari, K., & Hirabayashi, Y. (1996). Expression cloning of a cDNA for human ceramide glucosyltransferase that catalyzes the first glycosylation step of glycosphingolipid synthesis. *Proceedings of the National Academy of Sciences of the United States of America, 93*, 4638–4643.

Imachi, H., Nobu, M. K., Nakahara, N., Morono, Y., Ogawara, M., Takaki, Y., et al. (2020). Isolation of an archaeon at the prokaryote–eukaryote interface. *Nature, 577*, 519–525. https://doi.org/10.1038/s41586-019-1916-6.

Kentala, H., Weber-Boyvat, M., & Olkkonen, V. M. (2016). Chapter Seven OSBP-related protein family: Mediators of lipid transport and signaling at membrane contact sites. *International Review of Cell and Molecular Biology, 321*, 299–340. https://doi.org/10.1016/bs.ircmb.2015.09.006.

Kolesnick, R. N. (1987). 1,2-Diacylglycerols but not phorbol esters stimulate sphingomyelin hydrolysis in GH3 pituitary cells. *The Journal of Biological Chemistry, 262*, 16759–16762.

Kumagai, K., & Hanada, K. (2019). Structure, functions and regulation of CERT, a lipid-transfer protein for the delivery of ceramide at the ER–Golgi membrane contact sites. *FEBS Letters, 593*, 2366–2377. https://doi.org/10.1002/1873-3468.13511.

Levine, T. P., & Munro, S. (2002). Targeting of Golgi-specific pleckstrin homology domains involves both PtdIns 4-kinase-dependent and -independent components. *Current Biology, 12*, 695–704. https://doi.org/10.1016/s0960-9822(02)00779-0.

Limar, S., Körner, C., Martínez-Montañés, F., Stancheva, V. G., Wolf, V. N., Walter, S., et al. (2023). Yeast Svf1 binds ceramides and contributes to sphingolipid metabolism at the ER cis-Golgi interface. *Journal of Cell Biology, 222*, e202109162. https://doi.org/10.1083/jcb.202109162.

Lippincott-Schwartz, J., & Phair, R. D. (2010). Lipids and cholesterol as regulators of traffic in the endomembrane system. *Annual Review of Biophysics, 39*, 559–578. https://doi.org/10.1146/annurev.biophys.093008.131357.

Lipsky, N., & Pagano, R. (1985a). A vital stain for the Golgi apparatus. *Science (New York, N. Y.), 228*, 745–747. https://doi.org/10.1126/science.2581316.

Lipsky, N. G., & Pagano, R. E. (1983). Sphingolipid metabolism in cultured fibroblasts: Microscopic and biochemical studies employing a fluorescent ceramide analogue. *Proceedings of the National Academy of Sciences of the United States of America, 80*, 2608–2612.

Lipsky, N. G., & Pagano, R. E. (1985b). Intracellular translocation of fluorescent sphingolipids in cultured fibroblasts; endogenously synthesized sphingomyelin and glucosylcerebroside analogues pass through the Golgi apparatus en route to the plasma membrane. *The Journal of Cell Biology, 100*, 27–34.

Loewen, C. J. R., Roy, A., & Levine, T. P. (2003). A conserved ER targeting motif in three families of lipid binding proteins and in Opi1p binds VAP. *The EMBO Journal, 22*, 2025–2035. https://doi.org/10.1093/emboj/cdg201.

Lorente-Rodríguez, A., & Barlowe, C. (2011). Entry and exit mechanisms at the cis-face of the Golgi complex. *Cold Spring Harbor Perspectives in Biology, 3*, a005207. https://doi.org/10.1101/cshperspect.a005207.

Lorent, J. H., Levental, K. R., Ganesan, L., Rivera-Longsworth, G., Sezgin, E., Doktorova, M., et al. (2020). Plasma membranes are asymmetric in lipid unsaturation, packing and protein shape. *Nature Chemical Biology, 16*, 644–652. https://doi.org/10.1038/s41589-020-0529-6.

Lujan, P., & Campelo, F. (2021). Should I stay or should I go? Golgi membrane spatial organization for protein sorting and retention. *Archives of Biochemistry and Biophysics, 707*, 108921. https://doi.org/10.1016/j.abb.2021.108921.

Maceyka, M., & Spiegel, S. (2014). Sphingolipid metabolites in inflammatory disease. *Nature, 510*, 58–67. https://doi.org/10.1038/nature13475.

Maeda, K., Anand, K., Chiapparino, A., Kumar, A., Poletto, M., Kaksonen, M., et al. (2013). Interactome map uncovers phosphatidylserine transport by oxysterol-binding proteins. *Nature, 501*, 257–261. https://doi.org/10.1038/nature12430.

Mandon, E. C., Ehses, I., Rother, J., Echten, G. V., & Sandhoff, K. (1992). Subcellular localization and membrane topology of serine palmitoyltransferase, 3-dehydrosphinganine reductase, and sphinganine N-acyltransferase in mouse liver. *Journal of Biological Chemistry, 267*, 11144–11148. https://doi.org/10.1016/s0021-9258(19)49887-6.

Matteis, M. A. D., & Rega, L. R. (2015). Endoplasmic reticulum–Golgi complex membrane contact sites. *Current Opinion in Cell Biology, 35*, 43–50. https://doi.org/10.1016/j.ceb.2015.04.001.

Meng, E. C., Goddard, T. D., Pettersen, E. F., Couch, G. S., Pearson, Z. J., Morris, J. H., et al. (2023). UCSF ChimeraX: Tools for structure building and analysis. *Protein Science, 32*, e4792. https://doi.org/10.1002/pro.4792.

Mesmin, B., Bigay, J., Moser von Filseck, J., Lacas-Gervais, S., Drin, G., & Antonny, B. (2013). A four-step cycle driven by PI(4)P hydrolysis directs sterol/PI(4)P exchange by the ER-Golgi tether OSBP. *Cell, 155*, 830–843. https://doi.org/10.1016/j.cell.2013.09.056.

Mesmin, B., Kovacs, D., & D'Angelo, G. (2019). Lipid exchange and signaling at ER–Golgi contact sites. *Current Opinion in Cell Biology, 57*, 8–15. https://doi.org/10.1016/j.ceb.2018.10.002.

Michell, R. H. (1992). Inositol lipids in cellular signalling mechanisms. *Trends in Biochemical Sciences, 17*, 274–276. https://doi.org/10.1016/0968-0004(92)90433-a.

Mizuike, A., & Hanada, K. (2024). DGARM/C10orf76/ARMH3 for ceramide transfer zone at the endoplasmic reticulum–distal Golgi contacts. *Contact (Geneva, Switzerland), 7*, 25152564241239444. https://doi.org/10.1177/25152564241239443.

Mizuike, A., Sakai, S., Katoh, K., Yamaji, T., & Hanada, K. (2023). The C10orf76–PI4KB axis orchestrates CERT-mediated ceramide trafficking to the distal Golgi. *Journal of Cell Biology, 222*, e202111069. https://doi.org/10.1083/jcb.202111069.

Olkkonen, V. M., Johansson, M., Suchanek, M., Yan, D., Hynynen, R., Ehnholm, C., et al. (2006). The OSBP-related proteins (ORPs): Global sterol sensors for co-ordination of cellular lipid metabolism, membrane trafficking and signalling processes? *Biochemical Society Transactions, 34*, 389–391. https://doi.org/10.1042/bst0340389.

Pagano, R. E. (1988). What is the fate of diacylglycerol produced at the Golgi apparatus? *Trends in Biochemical Sciences, 13*, 202–205. https://doi.org/10.1016/0968-0004(88)90082-5.

Pagano, R. E., & Sleight, R. G. (1985). Defining lipid transport pathways in animal cells. *Science (New York, N. Y.), 229*, 1051–1057.

Posor, Y., Jang, W., & Haucke, V. (2022). Phosphoinositides as membrane organizers. *Nature Reviews Molecular Cell Biology, 23*, 797–816. https://doi.org/10.1038/s41580-022-00490-x.

Prinz, W. A., Toulmay, A., & Balla, T. (2020). The functional universe of membrane contact sites. *Nature Reviews Molecular Cell Biology, 21*, 7–24. https://doi.org/10.1038/s41580-019-0180-9.

Rodriguez-Gallardo, S., Kurokawa, K., Sabido-Bozo, S., Cortes-Gomez, A., Ikeda, A., Zoni, V., et al. (2020). Ceramide chain length–dependent protein sorting into selective endoplasmic reticulum exit sites. *Science Advances, 6*, eaba8237. https://doi.org/10.1126/sciadv.aba8237.

Rout, M. P., & Field, M. C. (2016). The evolution of organellar coat complexes and organization of the eukaryotic cell. *Annual Review of Biochemistry, 86*, 1–21. https://doi.org/10.1146/annurev-biochem-061516-044643.

Saito, S., Matsui, H., Kawano, M., Kumagai, K., Tomishige, N., Hanada, K., et al. (2008). Protein phosphatase 2Cϵ is an endoplasmic reticulum integral membrane protein that dephosphorylates the ceramide transport protein CERT to enhance its association with organelle membranes. *Journal of Biological Chemistry, 283*, 6584–6593. https://doi.org/10.1074/jbc.m707691200.

Santos, T. C. B., Dingjan, T., & Futerman, A. H. (2022). The sphingolipid anteome: Implications for evolution of the sphingolipid metabolic pathway. *FEBS Letters, 596*, 2345–2363. https://doi.org/10.1002/1873-3468.14457.

Shinoda, Y., Fujita, K., Saito, S., Matsui, H., Kanto, Y., Nagaura, Y., et al. (2012). Acyl-CoA binding domain containing 3 (ACBD3) recruits the protein phosphatase PPM1L to ER–Golgi membrane contact sites. *FEBS Letters, 586*, 3024–3029. https://doi.org/10.1016/j.febslet.2012.06.050.

Stankeviciute, G., Tang, P., Ashley, B., Chamberlain, J. D., Hansen, M. E. B., Coleman, A., et al. (2022). Convergent evolution of bacterial ceramide synthesis. *Nature Chemical Biology, 18*, 305–312. https://doi.org/10.1038/s41589-021-00948-7.

Tang, V. T., & Ginsburg, D. (2023). Cargo selection in endoplasmic reticulum–to–Golgi transport and relevant diseases. *Journal of Clinical Investigation, 133*, e163838. https://doi.org/10.1172/jci163838.

Thiele, I., Swainston, N., Fleming, R. M. T., Hoppe, A., Sahoo, S., Aurich, M. K., et al. (2013). A community-driven global reconstruction of human metabolism. *Nature Biotechnology, 31*, 419–425. https://doi.org/10.1038/nbt.2488.

Tidhar, R., & Futerman, A. H. (2013). The complexity of sphingolipid biosynthesis in the endoplasmic reticulum. *Biochimica et Biophysica Acta, 1833*, 2511–2518. https://doi.org/10.1016/j.bbamcr.2013.04.010.

Tóth, B., Balla, A., Ma, H., Knight, Z. A., Shokat, K. M., & Balla, T. (2006). Phosphatidylinositol 4-kinase IIIβ regulates the transport of ceramide between the endoplasmic reticulum and Golgi. *Journal of Biological Chemistry, 281*, 36369–36377. https://doi.org/10.1074/jbc.m604935200.

Vacaru, A. M., Tafesse, F. G., Ternes, P., Kondylis, V., Hermansson, M., Brouwers, J. F. H. M., et al. (2009). Sphingomyelin synthase-related protein SMSr controls ceramide homeostasis in the ER. *Journal of Cell Biology, 185*, 1013–1027. https://doi.org/10.1083/jcb.200903152.

Venditti, R., Masone, M. C., Rega, L. R., Tullio, G. D., Santoro, M., Polishchuk, E., et al. (2019a). The activity of Sac1 across ER–TGN contact sites requires the four-phosphate-adaptor-protein-1. *Journal of Cell Biology, 218*, 783–797. https://doi.org/10.1083/jcb.201812021.

Venditti, R., Rega, L. R., Masone, M. C., Santoro, M., Polishchuk, E., Sarnataro, D., et al. (2019b). Molecular determinants of ER–Golgi contacts identified through a new

FRET–FLIM system. *Journal of Cell Biology, 218*, 1055–1065. https://doi.org/10.1083/jcb.201812020.
Villani, M., Subathra, M., Im, Y.-B., Choi, Y., Signorelli, P., Del Poeta, M., et al. (2008). Sphingomyelin synthases regulate production of diacylglycerol at the Golgi. *Biochemical Journal, 414*, 31–41. https://doi.org/10.1042/bj20071240.
Weber, P., Hornjik, M., Olayioye, M. A., Hausser, A., & Radde, N. E. (2015). A computational model of PKD and CERT interactions at the trans-Golgi network of mammalian cells. *BMC Systems Biology, 9*, 9. https://doi.org/10.1186/s12918-015-0147-1.
Weigel, A. V., Chang, C.-L., Shtengel, G., Xu, C. S., Hoffman, D. P., Freeman, M., et al. (2021). ER-to-Golgi protein delivery through an interwoven, tubular network extending from ER. *Cell, 184*, 2412–2429.e16. https://doi.org/10.1016/j.cell.2021.03.035.
Yard, B. A., Carter, L. G., Johnson, K. A., Overton, I. M., Dorward, M., Liu, H., et al. (2007). The structure of serine palmitoyltransferase; gateway to sphingolipid biosynthesis. *Journal of Molecular Biology, 370*, 870–886. https://doi.org/10.1016/j.jmb.2007.04.086.
Zelnik, I. D., Ventura, A. E., Kim, J. L., Silva, L. C., & Futerman, A. H. (2020). The role of ceramide in regulating endoplasmic reticulum function. *Biochimica et Biophysica Acta – Molecular and Cell Biology of Lipids, 1865*, 158489. https://doi.org/10.1016/j.bbalip.2019.06.015.
Zhang, H., Desai, N. N., Olivera, A., Seki, T., Brooker, G., & Spiegel, S. (1991). Sphingosine-1-phosphate, a novel lipid, involved in cellular proliferation. *The Journal of Cell Biology, 114*, 155–167.

CHAPTER FOUR

Impact of coat protein on evolution of ilarviruses

Ali Çelik[a,*] and Adyatma Irawan Santosa[b]
[a]Department of Plant Protection, Faculty of Agriculture, Bolu Abant İzzet Baysal University, Bolu, Türkiye
[b]Department of Plant Protection, Faculty of Agriculture, Universitas Gadjah Mada, Yogyakarta, Indonesia
*Corresponding author. e-mail address: alicelik@ibu.edu.tr

Contents

Abstract

The genomic sequences attributed to the coat protein play a pivotal role in the evolutionary trajectory of plant viruses. The coat protein region, particularly scrutinized in the genus of *Ilarvirus* phylogroups, actively shapes the regional and host-specific dispersion. Within this chapter, assorted insights pertaining to the roles undertaken by coat proteins of frequently encountered Ilarviruses in their evolutionary processes are consolidated. Nonetheless, it is discerned that the availability of genomic data for RNA1 and RNA2 remains markedly limited, impeding the provision of lucid elucidations in this domain. Hence, to comprehensively delineate the evolution of Ilarviruses, a requisite exists for supplementary nucleotide sequence data, with a particular emphasis on taxa that have received lesser attention in research endeavors.

Genus *Ilarvirus* belong to the family Bromoviridae together with other economically devastating plant viruses such as cucumber mosaic virus (CMV, *Cucumovirus*), alfalfa mosaic virus (AMV, *Alfamovirus*), and brome mosaic virus (BMV, *Bromovirus*). These viruses shared common characters of having tripartite segregated positive-sense RNA genomes with a total length of around 8 kbp. The three genome segments are enveloped in separate virions that morphologically could be different: bacilliform or spherical.

RNA1 and RNA2, the larger genome segments of ilarviruses each possesses a single open reading frame, ORF1 and ORF2, that encode replicases, P1 and P2 protein, respectively. P1 is a non-structural protein involved in the virus replication (Pallas, Aparicio, Herranz, Sanchez-Navarro, & Scott, 2013). Similarly, P2 is an RNA-dependent RNA

Current Topics in Membranes, Volume 93
ISSN 1063-5823, https://doi.org/10.1016/bs.ctm.2024.05.002

polymerase (RdRp), one of elements of the replication process (Kozieł, Bujarski, & Otulak, 2017). The 5′ end of third segment (RNA3) contains ORF3a that encodes movement protein (MP) while ORF3b in the 3′ region encodes coat protein (CP) (Cui et al., 2015; Fajardo, Nascimento, Eiras, Nickel, & Pio-Ribeiro, 2015; Fiore et al., 2008; Kulshrestha et al., 2013). The MP is in the form of tubular structures which enable viral transportation from infected to their adjacent cells through tunnel-like plasmodesmata (Kozieł, Otulak-Kozieł, & Bujarski, 2018). The expression of CP is unique as it is started with transcription of a sub genomic region (sgRNA4) which promoter is located at ORF3a. In turn, sgRNA4 encodes ORF3b from which the viral coat protein (CP) is translated (Pallas et al., 2013). Beside encapsidating viral particles, CP of ilarviruses is also plays essential role in genome replication (Bachman, Scott, Xin, & Vance, 1994; Pallas et al., 2013).

There are 22 species currently approved by International Committee on Taxonomy of Viruses (ICTV) to be members of Ilarvirus. Most of the hosts are annual woody plants thus severe losses potentially occur during outbreaks. Thrips as vector, seeds, and pollen surface contamination are the main transmission modes from plant to plant. Infected plant materials and vegetative propagation through grafting are also crucial in long distance distribution. No wonder, some species: apple mosaic virus (ApMV), Prune dwarf virus (PDV), Prunus necrotic ringspot virus (PNRSV), and Tobacco streak virus (TSV) have occasionally caused large scale epidemics and may even global pandemic. Due to their importance, those species have been studied more extensively than other ilarviruses.

Numerous herbaceous and woody plants, and lichens worldwide have been naturally afflicted by the ApMV (Grimova, Winkowska, Zíka, & Ryšánek, 2016; Lakshmi et al., 2011; Petrzik & Lenz, 2002; Petrzik, Vondrák, Barták, Peksa, & Kubešová, 2014; Valasevich, Cieślińska, & Kolbanova, 2015). Despite extensive research, no definitive vector for ApMV has been identified (Postman & Mehlenbacher, 1992), though it is believed to be potentially transmitted through pollen and seed in hazelnut (Aramburu & Rovira, 2000). The persistence of this virus in vegetatively propagated material presents a significant epidemiological challenge, compounded by its easy transmission through grafting (Di Terlizzi, Digiaro, & Savino, 1991). Although ApMV may remain dormant within the host, plants affected by its infection display a range of symptoms, including leaf vein banding, yellow discoloration, and oak-leaf patterns (Nabi et al., 2022).

PDV is a prevalent pathogen affecting stone fruits (Paduch-Cichał, Sala-Rejczak, Mroczkowska, Boscia, & Potere, 2011; Vaskova, Petrzik, & Spak, 2000). Symptoms of PDV infection in plants manifest in various ways, such as deformity of leaves, yellowish rings and spots on leaves, and stunted growth. These symptoms can vary depending on factors such as the host plant, specific virus isolates, and environmental conditions (Kamenova, Borisova, & Popov, 2019; Predajňa et al., 2017). In addition to transmission through grafting, PDV can also be vertically transmitted via pollen or seeds (Kelley & Cameron, 1986; Silva, Tereso, Nolasco, & Oliveira, 2003), significantly impacting the health and yield of plants that are prone to infection (Uyemoto & Scott, 1992).

PNRSV is similar to PDV in terms of worldwide presence and stone fruits and apple as well as ornamental plants as hosts (Cui et al., 2015; Fiore et al., 2008; Verma, Hallan, Ram, & Zaidi, 2002; Çelik & Ertunç, 2019; Karanfil, 2021). The virus naturally spreads through pollen and seeds, with bees and thrips enhancing pollen transmission, and it can also disseminate through infected budwood and rootstock (Jones, 2018). Symptoms alone are unreliable indicators of PNRSV incidence; while some isolates may not manifest symptoms in infected *Prunus* plants, others may cause necrotic spots and shot holes on young leaves. In Bulgaria, PNRSV infection in sour and sweet cherry trees has been linked to necrotic ringspots with shot holes, accompanied by additional foliar symptoms such as chlorosis, leaf margin necrosis, deformation, and wrinkling (Kamenova & Borisova, 2021). Seasonal variations in symptoms have been observed, with chlorotic spots and mosaic patterns prevalent in spring, while shot holes and ringspots are more common later in the year (Sokhandan-Bashir, Kashiha, Koolivand, & Eini, 2017).

Genetic variation among isolates, phylogrouping, and evolution of genus *Ilarvirus* have been studied mainly through observation on the CP gene sequences of its three most successful species PNRSV, ApMV, and PDV. PNRSV is genetically separated into two major 'crown groups': PV96 and PV32, and several minor groups: CH30, SW6, PE5, and Pch Mx-Azt (Boulila, Tiba, & Jilani, 2013; Cui et al., 2015; Glasa, Betinová, Kúdela, & Šubr, 2002; Song et al., 2019; Xing et al., 2020). Isolates of the evolutionary successful PV96 and PV32 groups infecting wide range of hosts with distribution around the world, while those of the small groups tend to have limited regional presences. For instance, CH30 was detected in USA and UK (Glasa et al., 2002) while Pch Mx-Az was only detected in Mexico (Song et al., 2019) so far. There is no consensus yet on the naming of phylogroups of ApMV and PDV, other devastating ilarviruses on stone fruits, but recent population studies involving coat protein sequences of worldwide isolates registered in NCBI GenBank separated each virus into three main lineages and other noticeable point is that the MP and CP trees of each of ApMV and PDV were shown to have exact topography thus the same phylogroupings maybe observed using either MP or CP sequences (Çelik, Morca, Coşkan, & Santosa, 2023; Santosa, Çelik, Glasa, Ulubaş Serçe, & Ertunç, 2023).

Few traces of relationship between phylogroupings and geographical origins of isolates of those three viruses can still be observed. However, plant materials transportation through global trades are blurring them even more. Furthermore, detection works which were mostly carried out in certain regions only might not provide complete picture of isolates distribution. In ApMV, evidences for association of groups and hosts seem to be more tangible than location. The CP group 3 of ApMV phylogenetic tree is mostly dedicated to isolates from hop (*Humulus lupulus*) and hazelnut (*Corylus avellana*). None of hazelnut isolates was clustered in group 1 and 2, and only few hop isolates became members of group 1 (Çelik et al., 2023). Nevertheless, neither PNRSV nor PDV phylogeny showed similar correlation (Çelik, Santosa, Gibbs, & Ertunç, 2022; Santosa et al., 2023).

Besides clear naming consensus, characters of PNRSV phylogroups based on CP sequences were also better known than phylogroups of ApMV, PDV, and other ilarviruses. PV32 group is somewhat special as, biologically, its members were observed to be more virulent, causing more severe symptoms and dangerous diseases than isolates belonging to other groups (Hammond et al., 1999; Aparicio et al., 1999). For example, eight Turkish PV32 isolates (MT191363–70) were reported to naturally induce strong symptoms on peach leaves such as necrotic spots, chlorosis, malformation, and vein clearing (Çelik et al., 2022). PV32 isolates differ from isolates of other groups by having extra hexanucleotide (AAUAGG) which translate into two additional codons at position 41–42 in the CP of PNRSV. Although phylogenetically they are positioned in PV32 group, the eight Turkish peach isolates mentioned before did not have the extra hexanucleotide in their CP sequences. On the other hand, a very few isolates of other groups have them. Therefore, it can be concluded that the hexanucleotide might be eliminated during early divergence of PNRSV but gained again by PV32 variant before lost again in Turkish PV32 isolates.

Phylogenetic tree that was constructed in this article using alignment of complete sequences of coat protein region of 22 *Ilarvirus* species with Maximum Likelihood method based on Tamura 3-parameter's model

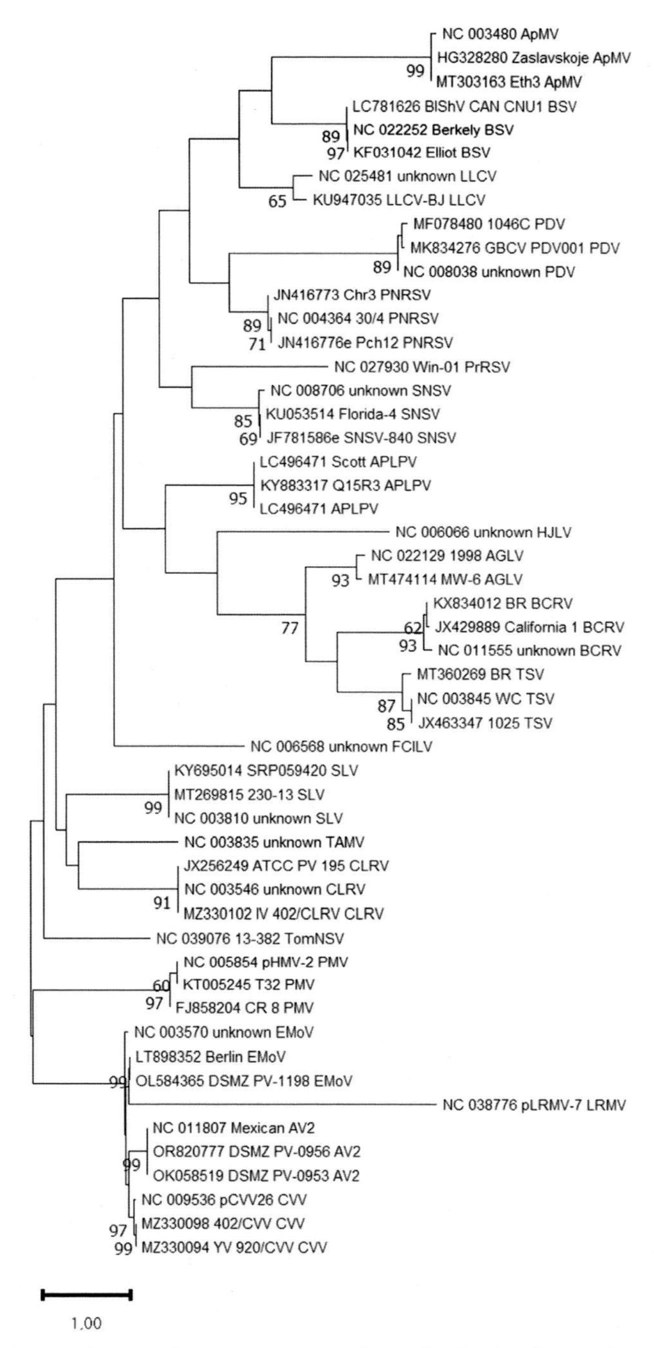

Fig. 1 Phylogenetic tree that was generated on the basis of complete nucleotide sequences of coat protein of 22 species of *Ilarvirus* using Maximum Likelihood

(Continued)

(Tamura, 1992) in MEGA 11 software (Tamura, Stecher, & Kumar, 2021) did not exhibit clear clustering of virus species and hosts. PDV and PNRSV, both infecting stone fruits around the world, were positioned together in a subcluster which in turn shared a basal node with another subcluster consists of ApMV, a pathogen also mostly infecting stone fruits, and found worldwide. TSV which infects hundreds of plant species, spread worldwide, and causing significant yield loss in warm regions was integrated in a subcluster distinct to subclusters of PDV, PNRSV, and ApMV (Fig. 1).

Worldwide isolates of ApMV, PDV, and PNRSV retained high nucleotide (nt) identities of around 85–100% at CP region and MP regions of ApMV and PDV were observed to have slightly higher nt identities than the CP regions (Çelik et al., 2022, 2023; Santosa et al., 2023).

Quantitative population analysis on the ApMV and PDV using DnaSP 6 software (Rozas et al., 2017) confirmed higher genetic diversity and polymorphisms on CP than MP sequences (Çelik et al., 2023; Santosa et al., 2023). The analysis involved different parameters including transcriptional selection ($\omega = dN/dS$) to estimate the variation among compared isolates. The observed gene is thought to be under positive (diversifying), neutral, or negative (purifying) selection when the respective ω value is >1, $=1$, or <1. In agreement with percentage identity results, CP regions of the two ilarviruses obtained higher S, η, and k values than the MP regions which suggested more divergence in CP than MP. These were probably due to currently more data available for CP than MP thus plenty CP variation can be examined.

The more vigorous selection pressure on MP than CP could be another reason for the few variations on MP. In PDV, the ω value assigned to CP region (0.179) was considerably higher than the one estimated for MP (0.067)

Fig. 1—Cont'd method based on Tamura 3-parameter's model (Tamura, 1992) with uniform rate among sites and complete deletion of missing data treatment implemented in MEGA 11 software. Significancy of branching was tested using 1000 bootstrap replicates, only >50% values were shown. Each species was represented by 1–3 isolates registered in NCBI GenBank. *Ageratum latent virus* (AGLV), *American plum line pattern virus* (APLPV), *Apple mosaic virus* (ApMV), *Asparagus virus 2* (AV2), *Blackberry chlorotic ringspot virus* (BCRV), *Blueberry shock virus* (BSV), *Citrus leaf rugose virus* (CLRV), *Citrus variegation virus* (CVV), *Elm mottle virus* (EMoV), *Fragaria chiloensis latent virus* (FCILV), *Humulus japonicus latent virus* (HJLV), *Lilac leaf chlorosis virus* (LLCV), *Lilac ring mottle virus* (LRMV), *Parietaria mottle virus* (PMV), *Privet ringspot virus* (PrRSV), *Prune dwarf virus* (PDV), *Prunus necrotic ringspot virus* (PNRSV), *Spinach latent virus* (SLV), *Strawberry necrotic shock virus* (SNSV), *Tobacco streak virus* (TSV), *Tomato necrotic streak virus* (TomNSV), *Tulare apple mosaic virus* (TAMV).

(Santosa et al., 2023). However, ω values for both CP and MP of ApMV were reported to be almost similar, 0.808 and 0.898, respectively (Çelik et al., 2023). CP of PNRSV was also given a low ω value of 0.177 (Çelik et al., 2022). Since the reported ω values were all below 1, MP and CP were shown to experience negative evolutionary constraints that keep variation in both genome regions stay limited.

Fu and Li's D^*, Fu and Li's F^*, and Tajima's D neutrality tests invariably estimated negative values to MP and CP of PNRSV, ApMV, and PDV populations (Çelik et al., 2022, 2023; Santosa et al., 2023). The three major ilarviruses were shown to experience population growth and expansion with lack of subdivision, or bottleneck selections thus they were indicated to have been in wide-scale epidemics. Besides that, new isolates capable to infect different host species may be assembled from this type of evolution. Pairwise comparisons of different ApMV and PDV phylogroups constantly produced Fixation index (F_{ST}) values > 0.25 which showed increasing genetic isolation between those phylogroups. The results also suggested that clustering isolates have been performed accordingly.

Observation on the three 'species model' of *Ilarvirus* showed relatively high genetic variation on the CP region thus the necessity to separate isolates into different phylogroups. The phylogroupings were found to be not strongly correlated to the origin of isolates or plant hosts. This suggested that the ability to infect certain plant species maybe related to mutations in other genomic regions. CP itself has been reported to involve in replication of Ilarvirus (Pallas et al., 2013). In the cases of PNRSV, ApMV, and PDV, although the MP and CP regions, both in RNA3, showed the same topographies, further analysis including RNA1 and RNA2 could not resolve phylogeny of each virus in the same manner (Çelik et al., 2022, 2023; Santosa et al., 2023). The lack of recombination in MP and CP regions (Kalinowska, Mroczkowska, Paduch-Cichal, & Chodorska, 2014) further indicated reassortment may have driven the evolution of Ilarvirus. Availability of genomic data of RNA1 and RNA2 are indeed still too low to provide clear pictures on this issue. Therefore, additional nucleotide sequence data, especially of the less studied species, are needed to fully explain the evolution of Ilarvirus.

References

Aparicio, F., Myrta, A., Terlizzi, B. D., & Pallás, V. (1999). Molecular variability among isolates of Prunus necrotic ringspot virus from different Prunus spp. *Phytopathology, 89*(11), 991–999. https://doi.org/10.1094/PHYTO.1999.89.11.99.

Aramburu, J., & Rovira, M. (2000). Incidence and natural spread of apple mosaic ilarvirus in hazel in north-east Spain. *Plant Pathology, 49*(4), 423–427.

Bachman, E. J., Scott, S. W., Xin, G. E., & Vance, V. B. (1994). The complete nucleotide sequence of prune dwarf ilarvirus RNA 3: Implications for coat protein activation of genome replication in ilarviruses. *Virology, 201*(1), 127–131. https://doi.org/10.1006/viro.1994.1272.

Boulila, M., Tiba, S. B., & Jilani, S. (2013). Molecular adaptation within the coat protein-encoding gene of Tunisian almond isolates of Prunus necrotic ringspot virus. *Journal of Genetics, 92*, 11–24. https://doi.org/10.1007/s12041-013-0211-9.

Çelik, A., & Ertunç, F. (2019). First report of prunus necrotic ringspot virus infecting apple in Turkey. *Journal of Plant Pathology, 101*(4), 1227.

Çelik, A., Morca, A. F., Coşkan, S., & Santosa, A. I. (2023). Global population structure of apple mosaic virus (ApMV, Genus Ilarvirus). *Viruses, 15*, 1221. https://doi.org/10.3390/v15061221.

Çelik, A., Santosa, A. I., Gibbs, A. J., & Ertunç, F. (2022). Prunus necrotic ringspot virus in Turkey: An immigrant population. *Archives of Virology, 167*, 553–562. https://doi.org/10.1007/s00705-022-05374-1.

Cui, H. G., Liu, H. Z., Chen, J., Zhou, J. F., Qu, L. N., Su, J. M., ... Hong, N. (2015). Genetic diversity of Prunus necrotic ringspot virus infecting stone fruit trees grown at seven regions in China and differentiation of three phylogroups by multiplex RT-PCR. *Crop Protection, 74*, 30–36. https://doi.org/10.1016/j.cropro.2015.04.001.

Di Terlizzi, B., Digiaro, M., & Savino, V. (1991). Ilaviruses in apricot and plum pollen. *In XV International Symposium on Fruit Tree Diseases, 309*, 93–98.

Fajardo, T. V. M., Nascimento, M. B., Eiras, M., Nickel, O., & Pio-Ribeiro, G. (2015). Molecular characterization of Prunus necrotic ringspot virus isolated from rose in Brazil. *Cienc Rural, 45*(12), 2197–2200. https://doi.org/10.1590/0103-8478cr20141810.

Fiore, N., Fajardo, T. V. M., Prodan, S., Herranz, M. C., Aparicio, F., Montealegre, J., ... Sánchez-Navarro, J. (2008). Genetic diversity of the movement and coat protein genes of South American isolates of Prunus necrotic ringspot virus. *Archives of Virology, 153*, 909–919. https://doi.org/10.1007/s00705-008-0066-1.

Glasa, M., Betinová, E., Kúdela, O., & Šubr, Z. (2002). Biological and molecular characterisation of Prunus necrotic ringspot virus isolates and possible approaches to their phylogenetic typing. *The Annals of Applied Biology, 140*, 279–283. https://doi.org/10.1111/j.1744-7348.2002.tb00182.x.

Grimova, L., Winkowska, L., Zíka, L., & Ryšánek, P. (2016). Distribution of viruses in old commercial and abandoned orchards and wild apple trees. *Journal of Plant Pathology*, 549–554.

Hammond, R. W., Crosslin, J. M., Pasini, R., Howell, W. E., & Mink, G. I. (1999). Differentiation of closely related but biologically distinct cherry isolates of Prunus necrotic ringspot virus by polymerase chain reaction. *Journal of Virological Methods, 80*(2), 203–212. https://doi.org/10.1016/S0166-0934(99)00051-8.

Jones, R. A. C. (2018). Plant and insect viruses in managed and natural environments: Novel and neglected transmission pathways. In C. M. Malmstrom (Ed.). *Advances in Virus Research: Environmental Virology and Virus Ecology* (pp. 149–187) Cambridge, Massachusetts, United States: Academic Press. https://doi.org/10.1016/bs.aivir.2018.02.006.

Kalinowska, E., Mroczkowska, K., Paduch-Cichal, E., & Chodorska, M. (2014). Genetic variability among coat protein of prune dwarf virus variants from different countries and different Prunus species. *European Journal of Plant Pathology, 140*(4), 863–868. https://doi.org/10.1007/s10658-014-0502-x.

Kamenova, I., & Borisova, A. (2021). Molecular variability of the coat protein gene of prunus necrotic ringspot virus on sweet and sour cherry in Bulgaria. *Journal of Plant Pathology, 103*, 97–104.

Kamenova, I., Borisova, A., & Popov, A. (2019). Incidence and genetic diversity of Prune dwarf virus in sweet and sour cherry in Bulgaria. *Biotechnology & Biotechnological Equipment, 33*(1), 980–987.

Karanfil, A. (2021). Prevalence and molecular characterization of Turkish isolates of the rose viruses. *Crop Protection, 143*, 105565.

Kelley, R. D., & Cameron, H. R. (1986). Associated with sweet cherry pollen and seed. *Phytopathology, 76*, 317–322.

Kozieł, E., Bujarski, J. J., & Otulak, K. (2017). Molecular biology of prune dwarf Virus—A lesser known member of the Bromoviridae but a vital component in the dynamic virus–host cell interaction network. *International Journal of Molecular Sciences, 18*(12), 2733. https://doi.org/10.3390/ijms18122733.

Kozieł, E., Otulak-Kozieł, K., & Bujarski, J. J. (2018). Ultrastructural analysis of prune dwarf virus intercellular transport and pathogenesis. *International Journal of Molecular Sciences, 19*(9), 2570. https://doi.org/10.3390/ijms19092570.

Kulshrestha, S., Hallan, V., Sharma, A., Seth, C. A., Chauhan, A., & Zaidi, A. A. (2013). Molecular characterization and intermolecular interaction of coat protein of Prunus necrotic ringspot virus: Implications for virus assembly. *Indian Journal of Virology, 24*(2), 235–241. https://doi.org/10.1007/s13337-013-0140-5.

Lakshmi, V., Hallan, V., Ram, R., Ahmed, N., Zaidi, A. A., & Varma, A. (2011). Diversity of Apple mosaic virus isolates in India based on coat protein and movement protein genes. *Indian Journal of Virology, 22*, 44–49.

Nabi, S. U., Baranwal, V. K., Rao, G. P., Mansoor, S., Vladulescu, C., Raja, W. H., ... Alansi, S. (2022). High-throughput RNA sequencing of mosaic infected and non-infected apple (Malus× domestica Borkh.) cultivars: From detection to the reconstruction of whole genome of viruses and viroid. *Plants, 11*(5), 675.

Paduch-Cichał, E., Sala-Rejczak, K., Mroczkowska, K., Boscia, D., & Potere, O. (2011). Serological characterization of Prune dwarf virus isolates. *Journal of Plant Protection Research, 51*(4), 389–392.

Pallas, V., Aparicio, F., Herranz, M. C., Sanchez-Navarro, J. A., & Scott, S. W. (2013). The molecular biology of ilarviruses. *Advances in Virus Research, 87*, 139–181. https://doi.org/10.1016/b978-0-12-407698-3.00005-3.

Petrzik, K., & Lenz, O. (2002). Remarkable variability of apple mosaic virus capsid protein gene after nucleotide position 141. *Archives of Virology, 147*, 1275–1285.

Petrzik, K., Vondrák, J., Barták, M., Peksa, O., & Kubešová, O. (2014). Lichens—A new source or yet unknown host of herbaceous plant viruses? *European Journal of Plant Pathology, 138*, 549–559.

Postman, J. D., & Mehlenbacher, S. A. (1992). Apple mosaic virus in hazelnut germplasm. *In III International Congress on Hazelnut, 351*, 601–610.

Predajňa, L., Sihelská, N., Benediková, D., Šoltys, K., Candresse, T., & Glasa, M. (2017). Molecular characterization of Prune dwarf virus cherry isolates from Slovakia shows their substantial variability and reveals recombination events in PDV RNA3. *European Journal of Plant Pathology, 147*, 877–885.

Rozas, J., Ferrer-Mata, A., Sánchez-DelBarrio, J. C., Guirao-Rico, S., Librado, P., Ramos-Onsins, S. E, & Sanchez-Gracia, A. (2017). DnaSP 6: DNA sequence polymorphism analysis of large data sets. *Molecular Biology and Evolution, 34*, 3299–3302. https://doi.org/10.1093/molbev/msx248.

Santosa, A. I., Çelik, A., Glasa, M., Ulubaş Serçe, Ç., & Ertunç, F. (2023). Molecular analysis of prune dwarf virus reveals divergence within non-Turkish and Turkish viral populations. *Journal of Plant Pathology, 105*(3), 943–954. https://doi.org/10.1007/s42161-023-01412-2.

Silva, C., Tereso, S., Nolasco, G., & Oliveira, M. M. (2003). Cellular location of Prune dwarf virus in almond sections by in situ reverse transcription-polymerase chain reaction. *Phytopathology, 93*(3), 278–285.

Sokhandan-Bashir, N., Kashiha, Z., Koolivand, D., & Eini, O. (2017). Detection and phylogenetic analysis of Prunus necrotic ringspot virus isolates from stone fruits in Iran. *Journal of Plant Pathology,* 723–729.

Song, S., Sun, P., Chen, Y., Ma, Q., Wang, X., Zhao, M., & Li, Z. (2019). Complete genome sequences of five prunus necrotic ringspot virus isolates from Inner Mongolia of China and comparison to other PNRSV isolates around the world. *Journal of Plant Pathology, 101*(4), 1047–1054. https://doi.org/10.1007/s42161-019-00335-1.

Tamura, K. (1992). Estimation of the number of nucleotide substitutions when there are strong transition-transversion and G + C-content biases. *Molecular Biology and Evolution, 9*(4), 678–687. https://doi.org/10.1093/oxfordjournals.molbev.a040752.

Tamura, K., Stecher, G., & Kumar, S. (2021). MEGA11: Molecular evolutionary genetics analysis version 11. *Molecular Biology and Evolution, 38*(7), 3022–3027. https://doi.org/10.1093/molbev/msab120.

Uyemoto, J. K., & Scott, S. W. (1992). Important diseases of prunus caused by viruses and other graft-transmissible pathogens in California and South Carolina. *Plant Disease, 76,* 5–11. https://doi.org/10.1094/PD-76-0005.

Valasevich, N., Cieślińska, M., & Kolbanova, E. (2015). Molecular characterization of Apple mosaic virus isolates from apple and rose. *European Journal of Plant Pathology, 141,* 839–845.

Vaskova, D., Petrzik, K., & Spak, J. (2000). Molecular variability of the capsid protein of the Prune dwarf virus. *The European Journal of Plant Pathology,* 106–573.

Verma, N., Hallan, V., Ram, R., & Zaidi, A. A. (2002). Detection of Prunus necrotic ringspot virus in begonia by RT-PCR. *Plant Pathology, 51*(6).

Xing, F., Gao, D., Liu, H., Wang, H., Habili, N., & Li, S. (2020). Molecular characterization and pathogenicity analysis of prunus necrotic ringspot virus isolates from China rose (*Rosa chinensis* Jacq.). *Archives of Virology, 165*(11), 2479–2486. https://doi.org/10.1007/s00705-020-04739-8.

CHAPTER FIVE

Lysosomal membrane contact sites: Integrative hubs for cellular communication and homeostasis

Sumit Bandyopadhyay, Daniel Adebayo, Eseiwi Obaseki, and Hanaa Hariri*

Department of Biological Sciences, Wayne State University, Detroit, MI, United States
*Corresponding author. e-mail address: hanaa.hariri@wayne.edu

Contents

Abstract

Lysosomes are more than just cellular recycling bins; they play a crucial role in regulating key cellular functions. Proper lysosomal function is essential for growth pathway regulation, cell proliferation, and metabolic homeostasis. Impaired lysosomal function is associated with lipid storage disorders and neurodegenerative diseases. Lysosomes form extensive and dynamic close contacts with the membranes of other organelles, including the endoplasmic reticulum, mitochondria, peroxisomes, and lipid droplets. These membrane contacts sites (MCSs) are vital for many lysosomal functions. In this chapter, we will explore lysosomal MCSs focusing on the machinery that mediates these contacts, how they are regulated, and their functional implications on physiology and pathology.

Current Topics in Membranes, Volume 93
ISSN 1063-5823, https://doi.org/10.1016/bs.ctm.2024.07.001

1. Introduction

Lysosomes, discovered by Christian de Duve in 1955, are membrane-enclosed organelles responsible for degrading proteins, lipids, carbohydrates, and nucleic acids (Ballabio, 2016; Trivedi et al., 2020; Sabatini & Adesnik, 2013). Mammalian cells contain several hundred lysosomes depending on the cell type and nutrient availability (Xu & Ren, 2015). In fungi, the equivalent of lysosomes are called vacuoles, with yeast cells housing generally one but up to 10 vacuoles of varying morphologies, that also change in response to nutrients and other signals (Chan & Marshall, 2014). Traditionally, lysosomes have been viewed as cellular recycling centers, responsible for breaking down various cellular components to maintain cellular homeostasis. The breakdown products are building blocks and metabolites for biosynthetic pathways.

Research over the past decade showed that lysosomes are more than just recycling bins. They act as signaling hubs by serving as a platform for the activation of mammalian target of rapamycin complex 1 mediating nutrient sensing and regulating cellular growth pathways. Additionally, lysosomes play a crucial role in regulating intracellular lipid homeostasis by interacting with other organelles and coordinating direct lipid transfer functions at these contact sites. Dysfunctional lysosomes are implicated in numerous diseases including neurodegenerative and metabolic disorders as well as cancers (Ballabio, 2016; Trivedi et al., 2020; Sabatini & Adesnik, 2013). Aberrant accumulation of biomolecules inside lysosomes is the hallmark of lysosomal storage disorders (LSDs), a group of rare genetic diseases which result in a range of symptoms including neurological issues.

Lysosomes are surrounded by a lipid bilayer studded with proteins and lined with a glycocalyx rich in oligosaccharides, which separates the acidic lumen of the lysosome (pH 4.5–5.0) from the cytosol. The lipid composition of the lysosomes is crucial for their functions. Lysosomes acquire their lipids either directly from other organelles via membrane contacts sites (MCSs) or from the endomembrane system via the endocytic pathway. A typical lysosomal membrane contains about 30% phosphatidylcholine, 11% phosphatidylethanolamine, and 7% phosphatidylinositol (Casares et al., 2019). Lysosomes typically contain a considerable amount of sphingolipids (15%) (Casares et al., 2019). Additionally, bis(monoacylglycero)phosphate (BMP), also known as lysobisphosphatidic acid, is unique lipid found on late endosomes and lysosomes and known to play important roles on the intraluminal vesicle membrane and promote lysosomal cholesterol egress (Gallala & Sandhoff, 2011; Medoh et al., 2022).

Numerous studies support the presence of segregated lipid microdomains in the lysosomal membranes enriched in sphingolipids and cholesterol, which regulates lysosomal functions. These microdomains are more prominent in the yeast lysosome/vacuole and have been implicated in regulating TORC1 signaling and lipid droplet (LD) turnover by lipophagy (Trivedi et al., 2020; Rudnik & Damme, 2021; Bouhamdani et al., 2021). Lysosomal membrane proteins are synthesized on the endoplasmic reticulum (ER) and targeted to lysosomes via the Golgi network. Over a hundred lysosomal transmembrane proteins have been identified and most of which are highly glycosylated to protect them from degradation by lysosomal hydrolases. Some of these proteins facilitate the movement of lipids across the lysosomal membrane (Rudnik & Damme, 2021; Schwake et al., 2013).

Many of the lysosomal roles are carried out through MCSs, where specialized proteins bring the membrane of lysosomes within 10–30 nm of other organelle membranes. Recent studies have shown that both lysosomal membrane proteins and lipids mediate inter-organelle contacts between the lysosome and other organelles, such as the ER, LDs, and mitochondria. This proximity allows for the direct transfer of lipids and ions, recruitment of growth regulators and signal molecules, and control of vesicle motility. In this chapter, we will discuss lysosomal contact sites with other organelles, what is known about their formation, regulation, and functional implications (Table 1) (Voeltz et al., 2024; Castro et al., 2022; Liao et al., 2022; Helle et al., 2013).

2. Discovery of the ER-lysosome/vacuole contact sites in yeast

MCSs proteins connecting the vacuole, the yeast counterpart of mammalian lysosomes, and the ER were first identified in yeast. Nvj1 was initially identified by Pan et al. (2000) through a yeast screen aimed at identifying interaction partners of Vac8, a yeast vacuolar membrane protein crucial for vacuole inheritance and cytoplasm-to-vacuole (Cvt) targeting. Together, these proteins (Nvj1-Vac8) form the nuclear ER-vacuole junction (NVJ).

Table 1 Known localization of proteins at lysosome membrane contact sites with their identified functions.

Protein (s)	Cellular location	Functions
Nvj1	Perinuclear ER	Piecemeal autophagy of the nucleus
Mdm1	Vacuole-nuclear ER-LD	Tethering, LD biogenesis, fatty acid metabolism, maintaining ER morphology
Snx13	Lysosome-ER-LD	Tethering, cholesterol homeostasis, prevention of cardiomyocyte apoptosis
Snx19	Lysosome-ER	Tethering, endolysosomal motility
Snx25	Lysosome-ER	Regulator of TGF-β signaling, autophagy
VPS13C	Lysosome-ER	Lipid transfer, lysosomal acidification
OSBP	Lysosome-ER	Tethering, mTORC1 pathway activation, regulator of lysosomal damage repair response
ORP1L	Lysosome-ER	Bidirectional cholesterol transfer, lysosome repair, regulator of autophagy
ORP5	Lysosome-ER	mTORC1 pathway activation, cell proliferation
SLC25A46	Lysosome-mitochondria	Mitochondrial cholesterol homeostasis, mitochondrial phospholipid homeostasis, mitochondrial morphology
Rab7-GTP	Lysosome-mitochondria	Contact formation, maintaining morphology of both lysosome and mitochondria
GDAP1 and MFN2	Lysosome-mitochondria	Formation and maintenance of contacts, lysosomal morphology, optimal autophagy
TRPML1	Lysosome-mitochondria	Mitochondrial Ca2+ homeostasis.
Rheb	Lysosome-Golgi	mTORC1 pathway activation
Syt7	Lysosome-peroxisome	Contact formation, cholesterol homeostasis
PLIN2	Lysosome-lipid droplet	Putative protein transfer

Nvj1 localizes exclusively to small patches on the perinuclear ER and is excluded from the cortical ER, even in Vac8Δ cells (Kvam & Goldfarb, 2006). Its N-terminus contains sequences necessary for the juxtaposition of the inner and outer nuclear membranes, reducing the distance between these membranes by ~9 nm compared to ~18 nm in the bulk nuclear envelope. This physical connection is essential for piecemeal autophagy of the nucleus (PMN), a process where both nuclear membranes and a fraction of the nucleoplasm are pinched off and degraded in the vacuole lumen (Millen et al., 2008). Deletion of either Nvj1 or Vac8 completely abrogates PMN (Roberts et al., 2003). It is known that PMN requires core autophagy-related (ATG) proteins (Krick et al., 2008); however, the exact mechanisms underlying the involvement of Nvj1 in PMN regulation remain unclear. Particularly whether this process is solely driven by Nvj1's physical linking of the inner and outer nuclear membrane and its interaction with Vac8, or if additional events are triggered downstream.

Under stress conditions, NVJ1 gene expression increases, leading to a proportional enlargement of the NVJ (Roberts et al., 2003). This concurrent increase in NVJ size, modulated by Nvj1 cellular levels, is likely mediated by stress-response elements in the NVJ1 promoter (Moskvina et al., 1998). The physiological impact for NVJ expansion on cellular homeostasis is not completely understood. However, research shows that, during stationary phase growth or in nutrient-limited conditions, numerous proteins are targeted to the NVJ region. These proteins are generally either lipid enzymes or lipid transfer proteins that are thought to compartmentalize lipid metabolism at the NVJ in response to nutrient depletion. One example is the oxysterol-binding protein Osh1 which, in log phase, localizes to both late-Golgi membranes and the NVJ. In the stationary phase, Osh1 associates exclusively with NVJ, and Nvj1 is necessary for osh1 recruitment to the NVJ (Kvam & Goldfarb, 2006; Levine & Munro, 2001). Osh1 contains lipid transfer domains and structural studies suggest that Osh1 performs non-vesicular transport of ergosterol and PI(4)P at the NVJ. In addition to Osh1, Nvj1 also recruits Tsc13 to the NVJ during nutrient starvation, where Tsc13 catalyzes the biosynthesis of very-long-chain fatty acids, integral components of various lipid species (Kvam et al., 2005). These lipid species are known constituents of detergent-insoluble lipid microdomains within cellular membranes (Dickson, 1998). In addition, recent work also revealed partitioning of HMG-CoA reductase at the NVJ in a Nvj1-dependent manner. This enzyme sub-compartmentalization enhances flux through the mevalonate pathway to ultimately promote sterol-ester synthesis during acute glucose restriction (Rogers et al., 2021). These findings, and others, suggest a role for the NVJ as a "metabolic platform" that coordinates the compartmentalization of enzymes in response to nutritional alterations in order to spatially regulate metabolism.

3. Functional protein tethers connect the ER and lysosomes

ER-lysosome contact sites have emerged as crucial cellular subdomains that facilitate direct communication between the ER and lysosomes to regulate various processes. Key proteins such as Oxysterol Binding Protein (OSBP), ORP1L, PI4K2A, VAPA/B, and NPC1 play pivotal roles at these contact sites. OSBP and ORP1L are involved in cholesterol sensing and bidirectional cholesterol transfer, respectively, while PI4K2A generates PI(4)P to recruit repair proteins following lysosomal membrane damage. Other proteins maintain lysosomal lipid homeostasis and facilitate lipid transfer at the contact sites. Proteins belonging to the sorting nexin–regulator of G-protein signaling (SNX-RGS) subfamily have also been shown to localize to the ER-lysosome contacts in yeast and mammalian cells. Snx13, which share homologies with yeast Mdm1 is implicated in ER-lysosome lipid transfer particularly cholesterol efflux, but the mechanisms remain unknown. Thus, ER-lysosome contact sites are crucial for cellular health and integrity, and they play various roles in facilitating lipid transfer, mTORC1 activation, lysosomal repair, and signal transduction. Understanding these interactions offers fundamental mechanistic insights into diseases associated with dysfunctional lysosomes. Below we will discuss the current knowledge of some of the key protein families that mediate lysosomal tethering to the ER and the processes they regulate at these sites. Below we explore the current knowledge about these proteins and their roles at the ER-lysosome contact sites.

3.1 SNX-RGS family of proteins

The SNX-RGS family of proteins is a subgroup of SNX superfamily, conserved from yeast to mammals. Yeast Mdm1 represents the SNX-RGS family, and it has four orthologs in mammals—Snx13, Snx14, Snx19, and Snx25. These proteins share a common structural organization including an

N-terminal transmembrane (TM) domain made with two hydrophobic alpha helices that anchor them to the ER (Henne et al., 2015). Except for Snx19, these proteins also contain an RGS domain (regulator of G-protein-mediated signaling). Additionally, all these proteins have a Phox homology (PX) domain that binds specific phosphoinositide species on the membranes of various organelles. This binding facilitates the formation of tethers between the ER and other organelles including the lysosome. The RGS and PX domains are flanked by PXA and PXC domains on their N-terminal and C-terminal sides, respectively (Paul et al., 2022; Hariri & Henne, 2022).

Currently, no experimental structures are available for these proteins. However, AlphaFold-predicted atomic structure analysis suggest that the PXA and PXC domains are intricately connected, exhibiting a similar arrangement across all members of this protein family (Fig. 1). The PXA and PXC domains are composed of α-helices and are intertwined such that neither domain is predicted to fold correctly without the other. Together, these domains form a tunnel-like structure approximately 80 Å in length. This tunnel's interior is lined with hydrophobic residues from both the PXA and PXC domains, and the highly conserved entry suggests functional significance (Paul et al., 2022; Hariri & Henne, 2022). We will discuss below the current knowledge of SNX-RGS proteins and their roles in cellular homeostasis in yeast and mammals.

3.1.1 Mdm1

Identified in a screen for proteins regulating endocytosis to the vacuole, Mdm1 is required for efficient vacuolar sorting of the endocytic reporter Mup1-pHluorin. Mdm1 localizes at sites of contact between the nuclear ER and the vacuole, also known as the NVJ. High resolution imaging revealed that Mdm1 localizes specifically to the rims of the NVJ patch and is excluded from the center suggesting a unique role for this protein tether at the NVJ. Mdm1 is anchored to the ER by two N-terminal integral membrane domains and to the vacuole by a C-terminal Phox (PX) domain bound to phosphatidylinositol-3-phosphate (PI3P). It localizes to the NVJ independently of Nvj1, and overexpression of Mdm1 causes hyper-tethering of the NVJ (Henne et al., 2015; Thomas et al., 2014a; Short, 2015; Datta et al., 2019).

Recent work revealed a role for Mdm1 in the biogenesis and turnover of lipid storage organelles LDs. Indeed, Mdm1 demarcates the sites of LDs budding at the rims of the NVJ. Overexpressing Mdm1 expands the NVJ

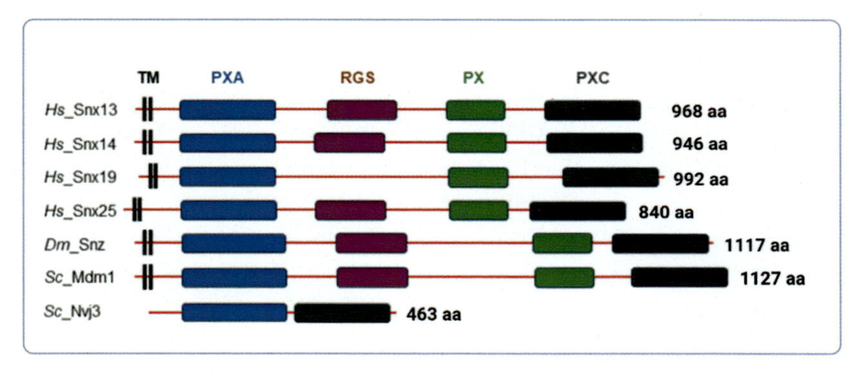

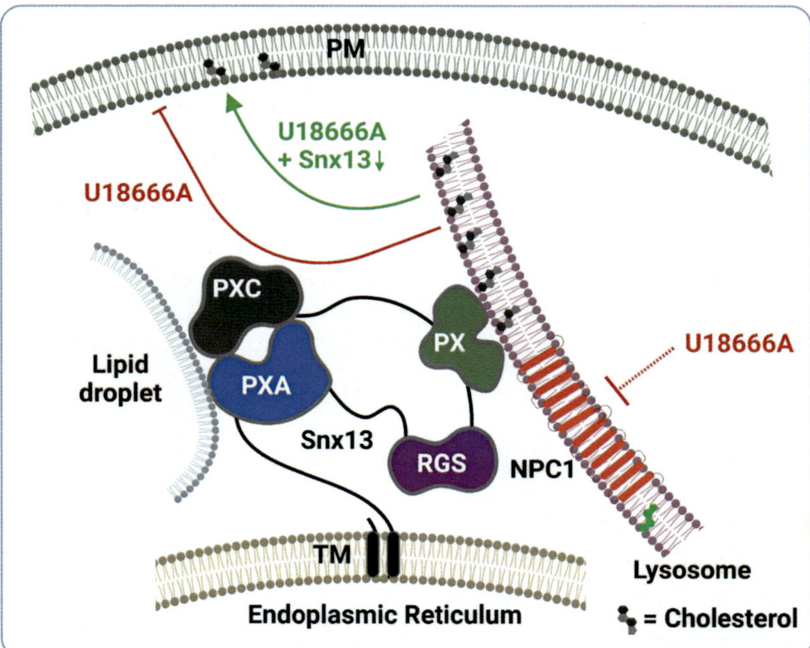

Fig. 1 Known localization of SNX-RGS family of interorganelle tethers. Upper panel: Humans (Homo sapiens, Hs) express four isoforms—Snx13, Snx14, Snx19, and Snx25. Fruit fly (Drosophila melanogaster, Dm) express one—Snz, and Yeast (*Saccharomyces cerevisiae*, Sc) express two—Mdm1 and Nvj3. Transmembrane (TM) domain anchors these proteins to the ER membrane. PX domain, except in Snx14, binds to phosphoinositide on endolysosome membrane. RGS domain regulated G-protein coupled signaling. PXA and PXC domains form a predicted lipid transfer cavity. Lower panel: proposed functional orientation of Snx13 bringing Endoplasmic Reticulum (ER), lysosome and Lipid droplet (LD) in close proximity and the PXA and PXC domains forming a predicted intertwined hydrophobic cavity. (1) Inhibition of NPC1 function by U18666A treatment results in the loss of lysosomal cholesterol egress and buildup of cholesterol on lysosomes. (2) However, downregulation of Snx13 expression in addition to NPC1 inhibition restores cholesterol egress from lysosomes to plasma membrane.

and drives LD accumulation of LDs. Thus, Mdm1 is thought to mediate a tri-organelle junction between the ER, vacuole, and LDs. Mdm1 interacts with Faa1, fatty acyl-CoA synthases presumably to facilitate fatty acids activation and incorporation into neutral lipids (triglycerides) that are typically stored in LDs. Consistent with this, data shows that the absence of Mdm1 leads to defects in the ER morphology resulting from toxic accumulation of fatty acids (Nguyen & Olzmann, 2019; Hariri et al., 2018, 2019; Obaseki et al., 2024). Loss-of-function of Mdm1 is also implicated in sphingolipid homeostasis and increases yeast cell sensitivity to myriocin, an inhibitor of ceramide synthesis. The molecular mechanism by which Mdm1 coordinates lipid synthesis and turnover at the NVJ, and their involvement in maintaining cellular lipid homeostasis remain an active area of research.

3.1.2 Snx13

Is the closest homologue to yeast Mdm1, and also functions as a tri-organelle tether, connecting the ER, lysosomes, and LDs (Fig. 2). Fluorescence imaging of Snx13 in U2OS cells showed that Snx13 redistributes to nascent LDs formed upon oleate supplementation. Additionally, Snx13 colocalized with the ER protein VAP-A, confirming its ER localization, and it colocalization with lysosomal marker LAMP1 was observed. Beyond its role as a multi-organelle tether, Snx13 regulates intracellular lipid homeostasis and is essential for healthy heart function (Li et al., 2014).

Snx13 was recently identified in a CRISPR screen as a regulator of lysosomal cholesterol (Lu et al., 2022). This study showed that inhibiting the lysosomal cholesterol exporter NPC1 led to expected cholesterol buildup in lysosomes. Surprisingly, this accumulation was reversed by the deletion of Snx13 suggesting that Snx13 is a negative regulator of lysosomal cholesterol export. Biochemical assays showed that siRNA depletion of Snx13 did not alter total cholesterol levels; however, Snx13 knockdown in U2OS and HeLa cells resulted in a redistribution of cholesterol from lysosomes to the plasma membrane. These data suggest that Snx13 mediates cholesterol efflux from the lysosome independent of the canonical NPC1 pathway (Lu et al., 2022). This study also confirmed the coregulation of cholesterol and BMP in cells where increased BMP levels have been shown to rescue endolysosomal cholesterol accumulation in NPC1-depleted cells (Chevallier et al., 2008; McCauliff et al., 2019). These findings suggest that, in the absence of NPC1 function, downregulation of Snx13 expression clears lysosomal cholesterol without

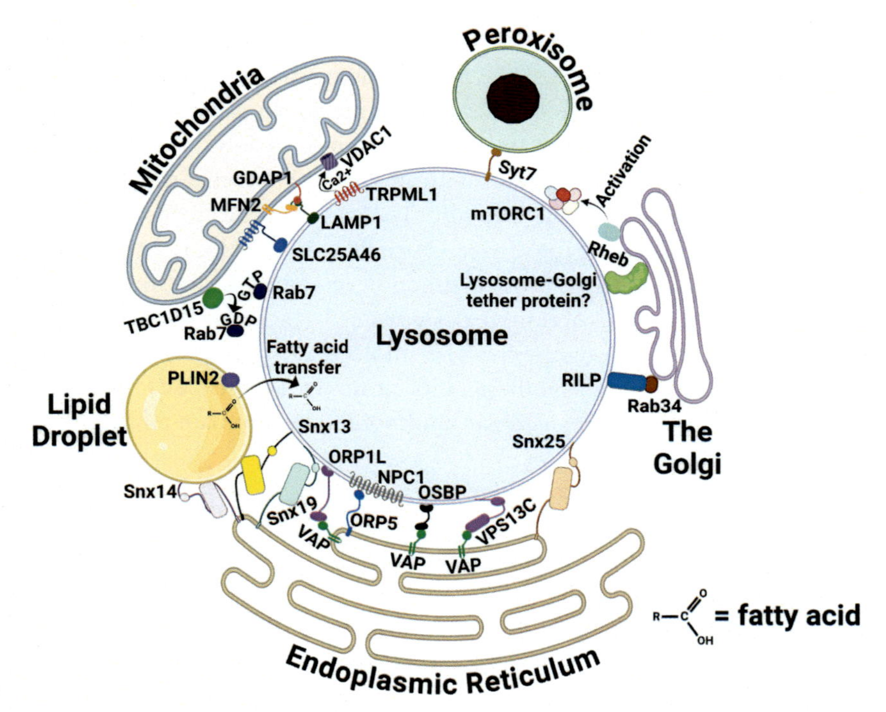

Fig. 2 Lysosome membrane contact site proteins discussed in the text. (1) Syt7 is a lysosomal membrane protein that forms contact with peroxisomes. (2) TRPML1 localizes at lysosome-mitochondria contact site on lysosome membrane and regulates mitochondrial Ca2+ homeostasis via mitochondria-localized VDAC1. (3) Mitochondria localized GDAP1 and MFN2 interacts with lysosome membrane protein LAMP1. (4) SLC25A46 is a mitochondrial outer membrane protein that localizes at mitochondria-lysosome MCS. (5) Rab7-GTP localizes on lysosome and can be hydrolyzed and released from lysosome by mitochondria localized GAP TBC1D15. (6) PLIN2 is located at the lysosome-LD contact site which is also the site for direct fatty acid transfer from LD to lysosome. (7) Snx13 is a triorganelle tether between LD-lysosome-ER. Snx19 and (8) Snx25 localize at lysosome-ER contact sites whereas Snx14 localizes at LD-ER contact sites. (9) ORP1L and OSBP are two ER-lysosome contact site proteins of the OSBP family that binds to the ER membrane protein VAP and lysosomal membrane. ORP5 of this protein family is an ER membrane anchored protein that interacts with NPC1. (10) VPS13C is another ER-lysosome MCS protein that interacts with ER membrane protein VAP. (11) Rheb localizes on the Golgi at Golgi-lysosome contact sites formed by unknown tether protein(s) and activates lysosome-localized mTORC1. (12) Golgi-localized Rab34 can interact with lysosome-localized RILP.

altering total cellular cholesterol, positioning Snx13 as a potential negative regulator of lysosomal cholesterol transport to the plasma membrane, with BMP likely supporting this clearance.

Reduced Snx13 protein levels was observed in failing hearts of humans and mice. This study was done in zebrafish where silencing of Snx13 caused heart dysfunction, indicating a vital role for Snx13 in heart health. Cardiomyocyte death was detected in the Snx13-deficient zebrafish heart, and siRNA knockdown of Snx13 resulted in apoptosis in rat ventricular cells. This study suggested a mechanism where downregulation of Snx13 activates apoptotic cell death, resulting in cardiomyocyte death and heart failure (Li et al., 2014). The results suggest a conserved role of Snx13 in heart cell survival. The mechanism by which Snx13 maintains cellular cholesterol homeostasis, and whether this is essential to protect heart function is not completely understood.

3.1.3 Snx19

Ubiquitously expressed protein that binds PI(3)P using its PX domain and also localizes to the ER-lysosome contact sites (Fig. 2). It colocalizes with the ER protein Calnexin, and this colocalization is abolished when the TM domain was truncated. Additionally, Snx19-GFP forms close appositions with LAMP1-RFP in live cells, supporting its role in lysosomal tethering. Besides lysosomes, Snx19 forms contacts with early and late endosomes. Snx19 tethering to endolysosomes is reduced by the R582Q mutation in its PX domain, which inhibits binding to PI3P. Interestingly, Snx19 lacking either its PXA or PXC domains forms "hypertethers" increasing the number of contacts and demonstrating the negative role of these domains in endolysosome tethering. CRISPR knockout of Snx19 in U2OS cells results in dispersed perinuclear ER-endolysosome contacts and exogenous expression of Snx19 rescues this phenotype. This shows that Snx19 is essential for proper ER-endolysosome contact formation. Furthermore, live-cell imaging in Snx19 knockout cells shows increased motility of endolysosomes, which is restricted in wild-type as well as in ΔPXA-Snx19 expressing cells, underscoring Snx19′s role in regulating endolysosomal motility (Saric et al., 2021).

3.1.4 Snx25

Is the third member of this protein family that localizes at ER-endolysosome contact sites (Fig. 2). Zebrafish Snx25 (zSnx25-PX) binds to PI(3)P in dimeric but not monomeric form, as determined by lipid overlay assay. Two mutations (R688E and K689E) in the PX domain abolished PI(3)P binding without affecting dimer formation, indicating that these residues are required for PI(3)P binding (Su et al., 2017). Both the PX and PXA domains of Snx25 are necessary for ER-endosome contact formation; deletion of either domain results in the loss of colocalization with the endosomal protein EEA1 (Early Endosome Antigen 1).

Studies in Hela cells suggest a role for Snx25 in regulating autophagy. Snx25 depletion increases the number of LC3 puncta, a marker of autophagy, suggesting either an increase in autophagosome synthesis or a decrease in autophagosome fusion and degradation via the lysosome. However, inhibition of autophagosome degradation with bafilomycin A1 in Snx25-depleted cells does not further increase in LC3 levels, indicating that indeed Snx25 depletion impairs autophagosome degradation. How deletion of Snx25 impairs autophagosome degradation is not clear. However, increased LC3 phenotype in Snx25 knockout cells was found to be rescued by the addition of ethanolamine, suggesting perturbed phospholipid metabolism in Snx25 knockout cells (Lauzier et al., 2022).

Other studies on Snx25 revealed a role in regulating TGF-β signaling. These studies showed that myc-tagged Snx25 interacts with TGF-β receptor I (TβRI) in HeLa cells. A luciferase activity assay revealed that Snx25 downregulates the TGF-β signaling pathway by reducing the levels of TβRI protein. Disruption of lysosomal activity by increasing its pH restores TβRI protein levels in 293 T cells, indicating that Snx25 likely mediates TβRI degradation through the lysosomal pathway. This underscores Snx25's role in regulating endolysosomal sorting and lysosomal degradation of TβRI thus regulating signaling by TGF-β (Hao et al., 2011).

Snx25 is the closest homologue of drosophila Snz (Snazurus) which unlike Snx25 does not localize to ER-endolysosome contacts. Snz has been implicated in organismal aging via an unknown mechanism (Suh et al., 2008). Snz is highly expressed in Drosophila fat body tissue where it localizes to ER-plasma MCSs and interacts with LDs at this location thus promoting a tri-organelle contact site (Hariri & Henne, 2022; Ugrankar et al., 2019). Like Snz, Snx25 has also been shown to associate with LDs following oleate treatment, but its full role in mediating LD autophagic flux remains to be determined (Paul et al., 2022; Lauzier et al., 2022).

3.1.5 Snx14

Loss-of-function mutations in Snx14 is linked to autosomal recessive spinocerebellar ataxia 20 (SCAR20) (Akizu et al., 2015; Bryant et al., 2018). Unlike other SNX-RGS proteins that interact with lysosomes, Snx14 uniquely localizes to the ER-LD interface (Datta et al., 2019) (Fig. 2). Whether Snx14 facilitates direct lipid transfer between the ER and LDs is not currently known. In cells, Snx14 interacts with fatty acid desaturase SCD1 at the ER-LD contact sites. This functional interaction between Snx14 and SCD1 is important for maintaining the lipid saturation balance of cell membranes. SNX14KO cells and SCAR20 disease patient-derived cells are hypersensitive to saturated fatty acids treatment which causes lipotoxic cell death (Datta et al., 2020). Although, why mutation in SNX14 cause SCAR20 is still not understood, these studies suggest a role for SNX14 in protecting the cell from saturated fatty acid toxicity.

3.2 VPS13 family of proteins

These are large proteins, approximately 300–500 kDa, initially identified through genetic screening in yeast. Four human orthologs of VPS13—VPS13A, VPS13B, VPS13C, and VPS13D—localize to specific subcellular regions (Kumar et al., 2018; Dziurdzik & Conibear, 2021; Leonzino et al., 2021). Among these, VPS13C localizes at the contact site between lysosomes and ER (Fig. 2), as well as the lysosomes and LDs. Overexpression of VPS13C in Cos7 cells showed a large pool colocalized with lysosomes, some of which were LAMP1 positive. Confocal imaging of HeLa and Cos-7 cells confirmed VPS13C localization at ER-lysosome contact sites.

VPS13C contains a canonical FFAT motif that enables it to bind to the ER membrane protein VAP (Cai et al., 2022) using its cytosolic MSP (Major Sperm Protein) domain (Kumar et al., 2018; Furuita et al., 2021). The VAB (VPS13 adapter binding) domain, located near the C-terminus of VPS13C, can interact with a Proline-X-Proline motif found in endolysosomal and mitochondrial adapters (Bean et al., 2018). The C-terminal pleckstrin homology (PH) domain, purified from yeast, has been shown to bind PI(4,5)P2 in liposomes (De et al., 2017). Additionally, a VPS13C fragment encompassing its WD40 region was found to localize on lysosomes, suggesting that this domain is responsible for its lysosomal localization (Kumar et al., 2018). These observations imply that multiple domains are involved in the lysosomal targeting of VPS13C.

The AlphaFold prediction of VPS13C structure reveals a 29.3-nm-long rod with a β-sheet running along its entire length. This β-sheet forms the floor of a hydrophobic tunnel, likely functioning as a lipid transfer channel. A disordered region flanking this rod-like structure contains the VAP-binding FFAT motif (Cai et al., 2022). This prediction aligns with the crystallographic structure previously obtained for the N-terminal region of VPS13C. This region, comprising residues 1–1350 and purified from yeast, was found to bind phosphatidylcholine, copurified with the VPS13C N-terminal fragment expressed in Expi293F cells. Approximately ten glycerophospholipids were bound to each fragment, suggesting VPS13C can bind multiple lipid molecules simultaneously. Therefore, VPS13 may act as bridges that support bulk lipid transfer between organelles (Melia & Reinisch, 2022).

An in vitro lipid transfer assay using donor liposomes contained fluorophore-tagged lipids, and donor and acceptor liposomes could be tethered by adding of N-terminal VPS13C fragment. FRET using this system detected lipid transfer from donor to acceptor liposomes after adding the VPS13C N-terminal fragment. The crystal structure of VPS13C residues 1–335 from *Chaetomium thermophilum* at 3.0 Å revealed a scoop-like structure with a cavity that is lined with hydrophobic amino acid residues measuring 20 Å across, supports the binding of multiple lipids at once (Kumar et al., 2018). Consistent with this, cryo-electron tomography of cells overexpressing VPS13C identified rod-shaped densities bridging the ER-lysosome contact sites. The length and diameter of this rod concur with the AlphaFold prediction of the VPS13C structure (Cai et al., 2022). These results suggest that VPS13C is a lipid transfer protein capable of transferring multiple molecules of lipids at ER-lysosome contact sites. The molecular mechanisms and regulation of lipid transfer by VPS13 at MCSs is not completely understood.

In a VPS13C-depleted dopaminergic neuron model, live-cell imaging showed significantly larger lysosomes and increased lysosome-lysosome tethering duration. Additionally, an increase in ER-lysosome contact formation was observed without any change in contact duration. Additionally, VPS13C-depleted neurons displayed reduced lysosomal motility compared to control cells, with most lysosomes restricted to the perinuclear region. These findings highlight the regulatory role of VPS13C in maintaining lysosomal morphology, dynamics, and contacts with the ER. Coimmunoprecipitation experiments demonstrated that VPS13C preferentially binds with phosphorylated Rab10, a small GTPase, which localizes at lysosomes expressing LAMP1-GFP. Inhibition of LRRK2 kinase blocks Rab10 phosphorylation, leading to reduced VPS13C-Rab10 interaction. These observations suggest that VPS13C on the ER binds to phosphorylated Rab10 on lysosomes, and Rab10 phosphorylation is required for this interaction.

Finally, in VPS13C-depleted neurons, a decrease in the levels of lysosomal proteases cathepsin B and cathepsin D was detected. Perturbation in lysosomal acidification was also observed by lysotracker staining in live cells. These findings were confirmed by shRNA targeting VPS13C, underscoring the role of VPS13C in proper lysosomal acidification and function (Schröder et al., 2024).

3.3 OSBP family of proteins

OSBP was initially discovered as a protein that binds to 25-hydroxycholesterol and other oxysterols, but not cholesterol (Kandutsch & Chen, 1977; Kandutsch & Thompson, 1980). It is known to downregulate the mevalonate pathway of cholesterol synthesis (Taylor et al., 1984). Later research identified OSBP as part of an evolutionarily conserved family of proteins characterized by the presence of ligand-binding OSBP-Related Domain (ORD). In yeast, there are seven OSBP homologs (Osh1p-Osh7p), classified into four subfamilies based on their ORD sequences, which denote their lipid-binding specificity (Jiang et al., 1994; Beh & Rine, 2004). In humans, eleven OSBP-Related Proteins (ORPs) have been identified, classified into six subfamilies (Laitinen et al., 1999; Jaworski et al., 2001; Lehto et al., 2008). Five of these proteins (ORP1, ORP3, ORP4, ORP8, ORP9) are expressed in either a long or a short length, each with a unique intracellular localization. These multidomain proteins feature both lipid-binding and a lipid-transport domains, with some possessing a C-terminal transmembrane domain (Eisenreichova et al., 2023).

All members of this protein family have a conserved EQVSHHPP sequence in their ORD domain. The structure of the ORD domain has been resolved in different members of the ORP proteins. The ORD domain of ORP3 contains a core of incomplete beta-barrel composed of antiparallel beta-sheet of hydrophobic residues. This is supplemented by two-stranded beta-sheet and four alpha-helices from the flanking N-terminal subdomain. Four α-helices and two β-strands flanking C-terminal domain form a hairpin. This structure forms the lipid-binding tunnel, with an entry closed by an amphipathic α-helix acting as a lid (Tong et al., 2018). Similar structural features are also seen in the ORD domain of ORP8 (Eisenreichova et al., 2023).

Most ORP proteins contain an N-terminal PH domain that can bind to specific phosphoinositides, facilitating association with membranes of various organelles, including lysosomes (Lemmon et al., 2002; Lemmon, 2008). Additionally, OSBP and ORP proteins have an FFAT (two phenylalanines in acidic tract) motif that binds to the cytosolic MSP domain of the ER-membrane protein VAP-A (de la Mora et al., 2021), forming contacts with the ER. ORP5 and ORP8 also have a transmembrane domain that anchors these proteins to the ER (Ghai et al., 2017). Among the members of this protein family, OSBP, ORP1L, and ORP5 are localized to the ER-lysosomes contact sites.

OSBP localization at ER-lysosome contact sites has been confirmed by imaging studies (Lim et al., 2019) (Fig. 2). Orp1L binds to lysosomal Rab7 via its ankyrin repeat domains and colocalizes with lysosomal LAMP1 in transiently transfected HeLa cells (Boutry & Kim, 2021; Johansson et al., 2005). This localization has also been observed in MelJuSo cells, with several other studies confirming Orp1L's presence on lysosomes (Van der Kant et al., 2013; Wijdeven et al., 2016; Zhao & Ridgway, 2017; Radulovic et al., 2022). Imaging studies have shown ORP1L localizing on the ER in HeLa and MelJuSo cells where its signal merges with the ER-specific marker Calnexin. A mutation in the FFAT motif (D478A) of ORP1L results in the loss of ER-tethering, indicating the importance of this motif for ER localization. Additionally, the ER transmembrane protein ORP5 interacts with NPC1, and deletion of the transmembrane domain results in lysosomal enrichment of ORP5 (Du et al., 2011).

OSBP and its related proteins play crucial roles in cholesterol sensing, mTORC1 activation, lysosomal repair, and autophagosome positioning. Under low cholesterol conditions, mTORC1 remains in the cytosol, but cholesterol promotes its translocation to the lysosomal membrane, activating its kinase function. This process is disrupted in the OSBP-inhibited HEK293 cells. In the absence of NPC1, cholesterol accumulates in the lysosomal limiting membrane, as detected by D4H staining. OSBP knockdown prevents this cholesterol buildup, unlike the knockdown of other ORPs, highlighting the unique role of OSBP.

Further evidence of OSBP's role in mTORC1 activation comes from experiments showing that cells expressing OSBP mutants, which cannot localize at ER-lysosome contact sites, do not translocate mTORC1 to lysosomes upon cholesterol supplementation. This observation is supported by shRNA depletion of VAPA/B, the ER binding partner of OSBP, demonstrating that cholesterol-mediated mTORC1 activation requires OSBP at ER-lysosome contact sites (Lim et al., 2019). Additionally, OSBP is involved in lysosomal damage repair (Radulovic et al., 2022). Treatment with L-leucyl-L-leucine methyl ester (LLOMe) induces rapid recruitment of PI(4)P to lysosomal membranes to facilitate lysosomal repair, but this process is blocked in OSBP-depleted HeLa cells, leading to cell death.

ORP1L is another key player in lysosomal damage repair. It was observed that inducing lysosome damage in HeLa cells by LLOMe treatment resulted in a rapid increase in lysosomal PI(4)P and a later increase in lysosomal cholesterol. Live-cell imaging showed lysosomal recruitment of ORP1L after damage induction and this recruitment was reduced under PI4K2A inhibition where cells were unable to accumulate PI(4)P on lysosomes. Furthermore, lysosomal repair, determined by lysotracker staining of damaged HeLa cells, was impaired when PI4K2A was inhibited. Finally, live-cell imaging showed that slow lysosomal accumulation of cholesterol in response to damage was abolished in cells where mutant ORP1L could not form ER-lysosome tethers. Intriguingly, in cells enriched with lysosomal cholesterol by U18666A treatment, LLOMe treatment induced weaker lysosomal damage. Taken together these results show that in response to damaged lysosomes, cells accumulate PI(4)P on lysosomes using PI4K2A which in turn leads to the recruitment of ORP1L at ER-lysosome contact sites followed by cholesterol-dependent lysosomal repair (Radulovic et al., 2022). While this work suggested that ORP1L may facilitate ER to lysosomes cholesterol transfer, another work has shown that ORP1L facilitates lysosomes to ER cholesterol transfer (Zhao & Ridgway, 2017). This work detected cholesterol buildup in lysosomes of ORP1L-null HeLa cells. This was accompanied by reduced cholesterol esterification in the ER suggesting that in the absence of ORP1L, cholesterol did not travel to the ER from lysosomes. Transient expression of exogenous ORP1L rescued these events supporting ORP1L as a cholesterol transfer protein from lysosomes to the ER. Additionally, an in vitro lipid binding assay demonstrated the ability of ORP1L to bind radiolabeled cholesterol. These observations established that ORP1L mediates lysosomal cholesterol egress (Zhao & Ridgway, 2017). Taken together, these works demonstrate that ORP1L can facilitate bidirectional cholesterol transfer at ER-lysosome MCS.

ORP1L also regulates autophagy (Wijdeven et al., 2016). Expression of a ΔORD mutant of ORP1L, unable to bind cholesterol, resulted in scattered distribution of autophagic vacuoles observed by confocal imaging and cryo-immuno-EM in HeLa and MelJuSo cells. Silencing of ORP1L reduces autophagic marker LC3 structures, and this decrease is reversed by blocking lysosomal degradation, indicating that ORP1L downregulates autophagic flux. ORP1L forms ER-autophagosome contacts, preventing dynein-dependent retrograde transport of autophagosomes under low cholesterol conditions (Wijdeven et al., 2016).

ORP5 is implicated in mTORC1 signaling. siRNA depletion of ORP5 downregulates mTORC1 pathway, indicated by decreased phosphorylation of mTOR substrate S6K in HeLa cells. ORP5 depletion also reduces translocation of mTORC1 complex to lysosomes. Coimmunoprecipitation shows mTOR interaction with ORP5, requiring the ORD domain of ORP1. Functional mutants of ORP5 with disrupted lipid binding exhibit reduced cell proliferation, suggesting that ORP5's lipid-binding promotes cell proliferation via mTORC1 activation (Du et al., 2018).

4. Lysosome-mitochondria contact sites

Lysosome-mitochondria contact sites are increasingly recognized as crucial for cellular homeostasis, particularly in coordinating metabolic and signaling pathways. These contact sites facilitate bidirectional communication, enabling the transfer of lipids, calcium, and other metabolites between lysosomes and mitochondria. Emerging research suggests that disruptions in lysosome-mitochondria contact sites are implicated in various diseases, including neurodegenerative disorders, highlighting their importance in maintaining cellular health and offering potential therapeutic targets. The dynamic nature of these contact sites allows for rapid response to cellular stress, linking lysosomal degradation capacity with mitochondrial energy production and apoptosis regulation. Proteins such as Rab7, VDAC1 (Voltage-Dependent Anion Channel 1), and TRPML1 play significant roles in these interactions, regulating processes like autophagy, energy metabolism, and apoptosis. Below we explore in detail the functions of some of these proteins.

4.1 Rab7

Is a small GTPase crucial for lysosomal biogenesis, localizing on lysosomes upon GTP binding (Bucci et al., 2000). Confocal imaging and electron microscopic analysis of HeLa cells confirmed Rab7′s presence on lysosomes (Bucci et al., 2000) (Fig. 2). In its GTP-bound state, Rab7 interacts with effector proteins on the membrane of other organelles and recruits membrane tethering factors (Cisneros et al., 2022). In healthy cells, Rab7-GTP facilitate dynamic lysosome-mitochondria contact sites, typically lasting for about 10 s. Expression of a constitutively active Rab7-GTP increases the number and duration of these contacts, suggesting that GTP hydrolysis reduces or abolishes them. However, these contacts do not result in mitophagy, indication a distinct role from damaged mitochondria to lysosomes for degradation (Wong et al., 2018).

TBC1D15, a GTPase activating protein (GAP), is recruited to mitochondria and hydrolyzes Rab7-GTP to Rab7-GDP (Yamano et al., 2014) (Fig. 2). A functionally inactive TBC1D15 mutant increases contact duration but not the formation of new contacts showing that GTP hydrolysis is necessary for untethering but not forming mitochondria-lysosome contacts. Proper localization of TBC1D15 to mitochondria is essential for untethering, as a mutants that fail to recruit TBC1D15 to mitochondria result in longer contacts. Functionally, mitochondria-lysosome contacts regulate organelle dynamics; mitochondrial fission sites colocalize with LAMP1-positive lysosomes before fission begins. Inhibiting GTP hydrolysis through constitutively active Rab7-GTP or inactive GAPs reduces mitochondrial fission and increases hyper-tethered and elongated mitochondria, along with abnormally large lysosomes. These findings suggest that Rab7-GTP regulates the dynamics and morphology of both mitochondria and lysosomes (Wong et al., 2018).

Recent studies revealed a role for Parkin, the E3 ubiquitin ligase that is mutated in Parkinson's disease, in regulating lysosome-mitochondria contact sites though maintaining active Rab7 bound on the lysosome (Peng et al., 2023). In iPSC-derived neurons from patients with Parkinson's disease, reduced mitochondria-lysosome contacts lead to accumulation of amino acids in lysosomes. Stabilizing these contacts in parkin-deficient neurons partially restores amino acid levels, emphasizing their role in maintaining compartment-specific amino acid profiles. Therefore, defects in Parkin contribute to Parkinsons's disease by disrupting mitochondria-lysosome contact dynamics and amino acid homeostasis, highlighting the therapeutic potential of targeting these contact sites to correct metabolic defects in disease.

4.2 GDAP1 and MFN2

Ganglioside-induced differentiation-associated protein 1 (GDAP1) is an integral, tail-anchored glutathione *S*-transferase located on the outer mitochondrial membrane (Huber et al., 2016; Cantarero et al., 2020) predominantly found in neurons (Fig. 2). GDAP1 directly interacts with the lysosomal membrane protein LAMP1, as confirmed by coimmunoprecipitation and proximity ligation assay (PLA) in SH-SY5Y cells. GDAP1-mediated mitochondria-lysosome contact sites formation is crucial for maintaining lysosomal morphology and optimal function. GDAP1 depletion led to a reduction in mitochondria-lysosome contact site surface, contact duration, and increased distance between these two organelles. Depletion of GDAP1 leads to a reduction in contact site surface area and duration, increased distance between these organelles, abnormally large lysosomes, and suboptimal basal autophagy. GDAP1 depletion also causes lower intracellular glutathione levels, essential for cellular redox homeostasis (Couto et al., 2016). Supplementing glutathione in GDAP1-depleted cells rescues abnormal lysosomal morphology and reduces the distance between mitochondria and lysosomes, though it does not restore optimal autophagy, indicating GDAP1's glutathione-independent role in autophagy maintenance (Cantarero et al., 2020).

Reduced mitochondria-lysosome contact formation has been observed in fibroblasts from individuals with missense mutation in their GDAP1 (W67L) (Pijuan et al., 2022), associated with neurological disorders. These cells exhibit an abnormally tangled and elongated mitochondrial network, reduced mitochondrial mass, and large lysosomal morphology, with an absence of autophagy in starvation. The study also identified another mitochondrial outer membrane protein, Mitofusin 2 (MFN2), a GTPase necessary for mitochondrial fusion, interacting with LAMP1 on the lysosomal membrane. Mutations in MFN2 (R104W) in patient fibroblasts further highlight the importance of GDAP1 and MFN2-mediated mitochondria-lysosome contact formation, essential for maintaining neurological health.

4.3 SLC25A46

(Solute carrier family 25 member 46) is an integral mitochondrial outer membrane protein that localizes at mitochondrial MCSs (Janer et al., 2016; Schuettpelz et al., 2023) (Fig. 2). Alkaline carbonate extraction of mitochondria from fibroblasts identified SLC25A46 in the outer membrane fraction alongside other outer membrane protein MFN2 (Janer et al., 2016). Live-cell imaging confirmed the presence of SLC25A46 at Mitochondria-Lysosomes contact sites, and a decrease in these interactions was observed in SLC25A46 knockout cell lines (Schuettpelz et al., 2023). In these SLC25A46 knockout fibroblasts, mitochondria exhibit reduced levels of free cholesterol without affecting the total cellular cholesterol. Additionally, a pulse-chase experiment using a fluorescent cholesterol probe revealed delayed cholesterol trafficking to the lysosomes in the knockout cells. Mitochondria in SLC25A46-deficient cells displayed a fragmented phenotype, which was rescued by cholesterol supplementation, indicating the proteins role in regulating cholesterol homeostasis (Janer et al., 2016).

During cellular starvation and stress, mitochondria undergoes stress-induced mitochondrial hyperfusion (SIMH), which confers a protective role. Lowered expression of SLC25A46 resulted in SIMH, but knockout cells failed to display this phenotype under starvation, reverting to SIMH only upon cholesterol addition. This suggests that SLC25A46 is essential for proper mitochondrial cholesterol regulation and stress response (Schuettpelz et al., 2023). Furthermore, phospholipids found in the mitochondrial membranes are mostly synthesized in the ER from where they can be transferred to mitochondria via mitochondria-associated ER membranes (Yeo et al., 2021). The phospholipid composition of the mitochondria was altered in cells from a patient carrying a homozygous missense mutation(C425CT) in SLC25A46 (Janer et al., 2016). These cells has reduced levels of PE, PA, and PS species and increased levels of PC plasmalogens along with abnormal sheet-like ER morphology, indicating a possible role for SLC25A46 in maintaining MCSs function and normal organelle morphology.

4.4 TRPML1

(transient receptor potential mucolipin 1) is a cation-permeable Ca2+ release channel located on the lysosomal membrane of all mammalian cells (Schmiege et al., 2017; Scotto Rosato et al., 2019) (Fig. 2). Loss-of-function mutations in TRPML1 cause LSDs and mitochondrial aberrations (Dong et al., 2008; Jennings et al., 2006; Eichelsdoerfer et al., 2010; Marques & Saftig, 2019). TRPML1 preferentially localizes to the mitochondria-lysosome contact sites and regulates their dynamics. Expression of a dominant-negative TRPML1 mutant increases the number and duration of these contact sites but decreases mitochondrial Ca2+ levels indicating TRPML1's role in Ca2+ transfer to mitochondria. Studies in HeLa cells, fibroblast, and HCT116 cells have shown that TRPML1 transfers Ca2+ from lysosomes to mitochondria at contact sites and this transfer does not occur to mitochondria that are not in contact with lysosomes. TBC1D15 knockout, which increases mitochondria-lysosome tethering, also raises mitochondrial Ca2+, further demonstrating the TRPML1-dependent Ca2+ transfer at these contact sites. Coimmunoprecipitation studies revealed TRPML1's interaction with VDAC1 present on the outer mitochondrial membrane, and VDAC1 mutant (E73Q) decreases mitochondrial Ca2+. Fibroblast from LSD patients with

TRPML1 mutations exhibit perturbed mitochondrial Ca2+ dynamics unlike those from healthy donors (Peng et al., 2020). Additionally, TRPML1 deletion in human natural killer cells results in mitochondrial fragmentation, reduced respiration, and ATP generation (Clement et al., 2023). These findings highlight the importance of TRPML1 at mitochondria-lysosome contact sites in maintaining normal organelle morphology, function, and Ca2+ homeostasis, linking it to LSDs. However, whether TRPML1 and VDAC1 act as tethers between mitochondria and lysosomes or are recruited to tethers formed by other factors remains unresolved.

5. Lysosome-Golgi contact sites

Despite their central role in the secretory pathway, contact sites with the Golgi apparatus remain largely unexplored (David et al., 2021). In mammalian cells, potential contacts between the Golgi and endosomes, as well as the Golgi and lysosomes, may be utilized in intracellular transport and signaling. Endocytosed sterols travel from lysosomes to the trans-Golgi network (TGN) both via vesicular and nonvesicular transport. This is facilitated by OSBP, RAB11, and RELCH, with the latter linking OSBP to RAB11 and mediating sterol transport from recycling endosomes to the TGN (Sobajima et al., 2018; Fujii et al., 2020). Depletion of these proteins results in reduced TGN sterol levels and increased late endosomal/lysosomal sterol accumulation. Additionally, Golgi-lysosome contacts form in response to amino acid stress, a condition that induces restricted lysosomal mobility and perinuclear clustering. This is mediated by RAB34, a GTPase mainly localized to the Golgi, and RILP, an effector of the tumor suppressor protein FLCN, which localizes to lysosomes (Starling et al., 2016).

Lysosome-Golgi contacts have also been suggested in amino acid repletion by Rheb proteins (Ras homolog enriched in brain) that localize to Golgi membranes and activate mTORC1 at lysosomes (Fig. 2). Rheb is a conserved G-protein of the Ras superfamily (Parmar & Tamanoi, 2010) and activates the master growth regulator mTORC1 (Yang et al., 2017). Despite it well-characterized function, Rheb's precise localization has been ambiguous, with reports indicating its presence on the ER (Hanker et al., 2010; Jiang & Vogt, 2008), lysosomes, Golgi (Thomas et al., 2014b; Manifava et al., 2016), and Peroxisomes (Zhang et al., 2013). Notably, both endogenous and transiently expressed Rheb localizes on Golgi across various cell types (Hao et al., 2018).

Under starvation conditions, mTORC1 remains inactive in the cytosol (Burnett et al., 1998; Jung et al., 2015). However, stable expression of Rheb-GFP is sufficient to activate lysosomal mTORC1 even during starvation (Hao et al., 2018). Targeting Rheb to lysosomes or the Golgi activates mTORC1, whereas ER-targeted of Rheb cannot. PLA using antibodies against Golgi protein GM130 and lysosomal protein LAMP1 detected contact spots between these organelles, with Rheb-GFP colocalizing at these sites. The absence of LAMP1 or GM130 eliminates these contacts, indicating their necessity for Golgi-lysosome contact formation. Disruption of these contacts using LLOMe treatment downregulates mTORC1 activation. Amino acid replenishment, which reduces mTORC1 translocating to lysosomes also increases Golgi-lysosome contacts mTORC1 activation (Ratto et al., 2022). These findings suggest that under abundant amino acid levels, mTORC1 translocates to lysosomes, which form more contact sites with the Golgi providing a platform for mTORC1 activation by Golgi-localized Rheb (Hao et al., 2018).

6. Lysosome-peroxisome contact sites

Studies have revealed that lysosome-peroxisome contact sites form in human cells, playing a crucial role beyond merely facilitating pexophagy, the degradation pathway for peroxisomes. These contact sites enable the direct transfer of cholesterol from lysosomes to peroxisomes. The tethering mechanism involves the integral lysosomal membrane protein Synaptotagmin VII (Syt7), which binds to PI(4,5)P2 on the peroxisomal membrane (Shai et al., 2016; Chen et al., 2020).

6.1 Syt7

Is a ubiquitously expressed Ca2+ sensor that regulates Ca2+-dependent exocytosis (Czibener et al., 2006). A screen using shRNA identified peroxisomal genes required for normal intracellular cholesterol trafficking, the depletion of these genes led to cholesterol buildup in lysosomes, indicating physical proximity between these organelles (Chu et al., 2015). Confocal imaging of HeLa cells showed contacts between peroxisomal PMP70 and lysosomal LAMP1, and electron microscopy of mouse liver cells confirmed more peroxisome-lysosome contacts. Time-lapse imaging revealed the dynamic nature of these contacts where one peroxisome remains in contact with a lysosome for up to 100 s. Refinement of the screen result revealed

Syt7 as a potential tether (Fig. 2), as Syt7 knockdown reduced lysosome-peroxisome contacts. Syt7 colocalized with LAMP1 confirming its lysosomal localization. Lipid-binding and liposome flotation assays showed that Syt7 binds $PI(4,5)P_2$ on peroxisomes. Depletion of $PI(4,5)P_2$ decreased lysosome-peroxisome contact formation and increased cholesterol accumulation, demonstrating that Syt7 on the lysosomes forms contact with PI (4,5)P2 on peroxisomes. Cholesterol tracking assays showed that cholesterol promotes contact site formation and it is transported from lysosomes to peroxisomes (Chu et al., 2015). Knockdown of PIP4K2A, the enzyme that converts PI5P to $PI(4,5)P_2$, reduced levels of $PI(4,5)P_2$ in peroxisomes, decreased contact formation, and led to cholesterol accumulation in lysosomes (Hu et al., 2018). In fibroblasts from patients with X-ALD, a neurodegenerative peroxisomal disorder (Engelen et al., 2014), significant intracellular cholesterol accumulation occurred, highlighting the role of peroxisomes in cholesterol homeostasis and its disruption in neurodegenerative disorder (Chu et al., 2015). Additionally, lysosomes serve as signaling hubs for mTOR, a master growth regulator. Peroxisome-localized TSC1/2 proteins repress mTOR signaling on the lysosome in response to oxidative stress suggesting a mechanism for coordinating mTOR signaling at lysosome-peroxisome contacts. This mechanism could be mediated via the Rheb GTPase-activating proteins in a manner similar to that proposed for the lysosome-Golgi contact sites (Zhang et al., 2013).

7. Lysosome-lipid droplets contact sites

Dynamic contacts between lysosomes and LD have been observed in AML12 hepatocytes and hepatocyte-derived cells using electron microscopy (Schulze et al., 2020). This research found an average of 30.26 s of interaction between BODIPY-labeled LDs and Lysotracker-labeled lysosomes under basal culture conditions. The interaction between LDs and lysosomes are thought to facilitate LD turnover. However, culturing these cells under nutrient starvation, a trigger for autophagy, did not increase these contacts duration. Interestingly, the depletion of macroautophagy and chaperone-mediated autophagy (CMA) proteins did not reveal major changes in these organelle contacts either thus suggesting that Lysosome-LD contacts are independent of macroautophagy and CMA. Transfection of a dual fluorescent PLIN2 results in yellow fluorescence when the PLIN2 protein localizes on LDS (Fig. 2) but its interaction with acidic compartments produces a red fluorescence. Long-term live-cell imaging using this system showed initial yellow fluorescence from LDs getting converted to red-only fluorescence punctae along with their dissociation from source LDs, suggesting a possibility of protein transfer from LDs to lysosomes. Notably, this suggested LD to lysosome protein transfer occurred even in nutrient-rich culture conditions. In addition to proteins, BODIPY-labeled C12 fatty acids were observed to accumulate in lysosomes surrounding LDs during contact formation. Electron microscopy of hepatocytes that were stimulated for LD formation and catabolism showed numerous instances of direct lipid transfer from LDs to lysosomes. This lipid transfer was not frequently observed in cells cultured in adequate nutrient media. Using a drug (Lalistat) that inhibits the function of lysosomal acid lipases resulted in the accumulation of neutral lipids in lysosomes. Intriguingly, inhibition of CMA and macroautophagy by knocking down LAMP2A and atg5, respectively, did not significantly decrease this accumulation. Taken together, this study highlights the possible existence of a lipid catabolism pathway that is independent of canonical autophagy and is mediated by lysosome-LD contact formation (Schulze et al., 2020).

Another study identified a role of Rab5 in lysosome-LD interactions (Schott et al., 2024). Confocal imaging showed disruption of lysosome-LD interactions in primary rat hepatocytes as well as in HepG2 cells under ethanol exposure. siRNA knockdown of Rab5, which was detected in purified LDs, resulted in an increase in both LD number as well as LD area per cell. Interestingly, expression of a dominant negative form of Rab5 resulted in abolished lysosome-LD interactions suggesting the importance of Rab5 GTPase activity in these interactions. Surprisingly, Rab5 enrichment on LDs was observed in the hepatocytes of rats that received a chronic ethanol diet. In addition, the ethanol diet did not alter the GTPase activity of Rab5. Ethanol inhibits the GTPase activity of Rab7, which works downstream of Rab5, and inhibiting the GTPase activity of Rab7 resulted in an accumulation of Rab5 on LDs. These observations suggest a possible pathway of Rab5-dependent lysosome-LD interactions that requires the GTPase activity of Rab7 and is disrupted by ethanol exposure (Schott et al., 2024).

8. Conclusions and perspectives

Research into the formation of contact sites between the lysosome and other organelles is rapidly evolving, with much remaining to be understood about the intricate molecular mechanisms involved, including tethering, recruitment of regulatory factors, and interplay among other tether protein and complexes. Nevertheless, significant progress has been made so far, shedding light on the crucial roles of lysosomes in cellular lipid and metabolite homeostasis, signaling pathways, cell proliferation, and the maintenance of overall organelle function. In essence, the formation of inter-organelle contacts by healthy lysosomes is paramount for ensuring multiple facets of cellular health and functionality. Dysfunction in lysosomal activity has been linked to metabolic disorders like LSDs and various neurodegenerative diseases. How lysosomal contact sites machinery contribute to the pathology of these disorders is currently an open area of research.

References

Akizu, N., et al. (2015). Biallelic mutations in SNX14 cause a syndromic form of cerebellar atrophy and lysosome-autophagosome dysfunction. *Nature Genetics, 47*, 528–534.

Ballabio, A. (2016). The awesome lysosome. *EMBO Molecular Medicine, 8*, 73–76.

Bean, B. D. M., et al. (2018). Competitive organelle-specific adaptors recruit Vps13 to membrane contact sites. *The Journal of Cell Biology, 217*, 3593–3607.

Beh, C. T., & Rine, J. (2004). A role for yeast oxysterol-binding protein homologs in endocytosis and in the maintenance of intracellular sterol-lipid distribution. *Journal of Cell Science, 117*, 2983–2996.

Bouhamdani, N., Comeau, D., & Turcotte, S. (2021). A compendium of information on the lysosome. *Frontiers in Cell and Developmental Biology, 9*.

Boutry, M., & Kim, P. K. (2021). ORP1L mediated PI(4)P signaling at ER-lysosome-mitochondrion three-way contact contributes to mitochondrial division. *Nature Communications, 12*(1), 5354. https://www.nature.com/articles/s41467-021-25621-4.

Bryant, D., et al. (2018). SNX14 mutations affect endoplasmic reticulum-associated neutral lipid metabolism in autosomal recessive spinocerebellar ataxia 20. *Human Molecular Genetics, 27*, 1927–1940.

Bucci, C., Thomsen, P., Nicoziani, P., McCarthy, J., & van Deurs, B. (2000). Rab7: A key to lysosome biogenesis. *Molecular Biology of the Cell, 11*(2), 413–793. https://www.molbiolcell.org/doi/10.1091/mbc.11.2.467.

Burnett, P. E., Barrow, R. K., Cohen, N. A., Snyder, S. H., & Sabatini, D. M. (1998). RAFT1 phosphorylation of the translational regulators p70 S6 kinase and 4E-BP1. *Proceedings of the National Academy of Sciences of the United States of America, 95*, 1432–1437.

Cai, S., et al. (2022). In situ architecture of the lipid transport protein VPS13C at ER–lysosome membrane contacts. *Proceedings of the National Academy of Sciences of the United States of America, 119*, e2203769119.

Cantarero, L., Juárez-Escoto, E., Civera-Tregón, A., Rodríguez-Sanz, M., Roldán, M., Benítez, R., et al. (2020). Mitochondria–lysosome membrane contacts are defective in GDAP1-related Charcot–Marie–Tooth disease. *Human Molecular Genetics, 29*(22), 3589–3605. https://academic.oup.com/hmg/article/29/22/3589/5958071?login=true.

Casares, D., Escribá, P. V., & Rosselló, C. A. (2019). Membrane lipid composition: Effect on membrane and organelle structure, function and compartmentalization and therapeutic avenues. *International Journal of Molecular Sciences, 20*, 2167.

Castro, I. G., et al. (2022). Systematic analysis of membrane contact sites in *Saccharomyces cerevisiae* uncovers modulators of cellular lipid distribution. *eLife, 11*, e74602.

Chan, Y.-H. M., & Marshall, W. F. (2014). Organelle size scaling of the budding yeast vacuole is tuned by membrane trafficking rates. *Biophysical Journal, 106*, 1986–1996.
Chen, C., Li, J., Qin, X., & Wang, W. (2020). Peroxisomal membrane contact sites in mammalian cells. *Frontiers in Cell and Developmental Biology, 8*.
Chevallier, J., et al. (2008). Lysobisphosphatidic acid controls endosomal cholesterol levels. *Journal of Biological Chemistry, 283*, 27871–27880.
Chu, B.-B., et al. (2015). Cholesterol transport through lysosome-peroxisome membrane contacts. *Cell, 161*, 291–306.
Cisneros, J., Belton, T. B., Shum, G. C., Molakal, C. G., & Wong, Y. C. (2022). Mitochondria-lysosome contact site dynamics and misregulation in neurodegenerative diseases. *Trends in Neurosciences, 45*, 312–322.
Clement, D., Szabo, E. K., Krokeide, S. Z., Wiiger, M. T., Vincenti, M., Palacios, D., et al. (2023). The lysosomal calcium channel TRPML1 maintains mitochondrial fitness in NK cells through interorganelle cross-talk. *The Journal of Immunology, 211*(9), 1348–1358. https://journals.aai.org/jimmunol/article/211/9/1348/265924/The-Lysosomal-Calcium-Channel-TRPML1-Maintains.
Couto, N., Wood, J., & Barber, J. (2016). The role of glutathione reductase and related enzymes on cellular redox homoeostasis network. *Free Radical Biology and Medicine, 95*, 27–42. https://www.sciencedirect.com/science/article/pii/S0891584916000873?via%3Dihub.
Czibener, C., et al. (2006). Ca2+ and synaptotagmin VII–dependent delivery of lysosomal membrane to nascent phagosomes. *Journal of Cell Biology, 174*, 997–1007.
Datta, S., et al. (2020). Snx14 proximity labeling reveals a role in saturated fatty acid metabolism and ER homeostasis defective in SCAR20 disease. *Proceedings of the National Academy of Sciences of the United States of America, 117*, 33282–33294.
Datta, S., Liu, Y., Hariri, H., Bowerman, J., & Henne, W. M. (2019). Cerebellar ataxia disease–associated Snx14 promotes lipid droplet growth at ER–droplet contacts. *Journal of Cell Biology, 218*, 1335–1351.
David, Y., Castro, I. G., & Schuldiner, M. (2021). The fast and the Furious: Golgi contact sites. *Contact (Thousand Oaks), 4*, 25152564211034424.
de la Mora, E., et al. (2021). Nanoscale architecture of a VAP-A-OSBP tethering complex at membrane contact sites. *Nature Communications, 12*, 3459.
De, M., et al. (2017). The Vps13p-Cdc31p complex is directly required for TGN late endosome transport and TGN homotypic fusion. *The Journal of Cell Biology, 216*, 425–439.
Dickson, R. C. (1998). Sphingolipid functions in *Saccharomyces cerevisiae*: Comparison to mammals. *Annual Review of Biochemistry, 67*, 27–48.
Dong, X.-P., et al. (2008). The type IV mucolipidosis-associated protein TRPML1 is an endolysosomal iron release channel. *Nature, 455*, 992–996.
Du, X., et al. (2018). Oxysterol-binding protein–related protein 5 (ORP5) promotes cell proliferation by activation of mTORC1 signaling. *Journal of Biological Chemistry, 293*, 3806–3818.
Du, X., Kumar, J., Ferguson, C., Schulz, T. A., Ong, Y. S., Hong, W., et al. (2011). A role for oxysterol-binding protein-related protein 5 in endosomal cholesterol trafficking. *Journal of Cell Biology, 192*, 121–135. https://rupress.org/jcb/article/192/1/121/36300/A-role-for-oxysterol-binding-protein-related.
Dziurdzik, S. K., & Conibear, E. (2021). The Vps13 family of lipid transporters and its role at membrane contact sites. *International Journal of Molecular Sciences, 22*, 2905.
Eichelsdoerfer, J. L., Evans, J. A., Slaugenhaupt, S. A., & Cuajungco, M. P. (2010). Zinc dyshomeostasis is linked with the loss of mucolipidosis IV-associated TRPML1 ion channel. *The Journal of Biological Chemistry, 285*, 34304–34308.
Eisenreichova, A., et al. (2023). Crystal structure of the ORP8 lipid transport ORD domain: Model of lipid transport. *Cells, 12*, 1974.

Engelen, M., Kemp, S., & Poll-The, B.-T. (2014). X-Linked adrenoleukodystrophy: Pathogenesis and treatment. *Current Neurology and Neuroscience Reports, 14*, 486.

Fujii, S., et al. (2020). Recycling endosomes attach to the trans-side of Golgi stacks in Drosophila and mammalian cells. *Journal of Cell Science, 133*, jcs236935.

Furuita, K., Hiraoka, M., Hanada, K., Fujiwara, T., & Kojima, C. (2021). Sequence requirements of the FFAT-like motif for specific binding to VAP-A are revealed by NMR. *FEBS Letters, 595*, 2248–2256.

Gallala, H. D., & Sandhoff, K. (2011). Biological function of the cellular lipid BMP—BMP as a key activator for cholesterol sorting and membrane digestion. *Neurochemical Research, 36*, 1594–1600.

Ghai, R., et al. (2017). ORP5 and ORP8 bind phosphatidylinositol-4, 5-biphosphate (PtdIns(4,5)P2) and regulate its level at the plasma membrane. *Nature Communications, 8*, 757.

Hanker, A. B., et al. (2010). Differential requirement of CAAX-mediated posttranslational processing for Rheb localization and signaling. *Oncogene, 29*, 380–391.

Hao, F., et al. (2018). Rheb localized on the Golgi membrane activates lysosome-localized mTORC1 at the Golgi–lysosome contact site. *Journal of Cell Science, 131*, jcs208017.

Hao, X., et al. (2011). SNX25 regulates TGF-β signaling by enhancing the receptor degradation. *Cellular Signalling, 23*, 935–946.

Hariri, H., & Henne, W. M. (2022). Filling in the gaps: SNX-RGS proteins as multi-organelle tethers. *Journal of Cell Biology, 221*, e202203061.

Hariri, H., et al. (2018). Lipid droplet biogenesis is spatially coordinated at ER-vacuole contacts under nutritional stress. *EMBO Reports, 19*, 57–72.

Hariri, H., et al. (2019). Mdm1 maintains endoplasmic reticulum homeostasis by spatially regulating lipid droplet biogenesis. *Journal of Cell Biology, 218*, 1319–1334.

Helle, S. C. J., et al. (2013). Organization and function of membrane contact sites. *Biochimica et Biophysica Acta (BBA) – Molecular Cell Research, 1833*, 2526–2541.

Henne, W. M., et al. (2015). Mdm1/Snx13 is a novel ER-endolysosomal interorganelle tethering protein. *The Journal of Cell Biology, 210*, 541–551.

Hu, A., et al. (2018). PIP4K2A regulates intracellular cholesterol transport through modulating PI(4,5)P2 homeostasis. *Journal of Lipid Research, 59*, 507–514.

Huber, N., et al. (2016). Glutathione-conjugating and membrane-remodeling activity of GDAP1 relies on amphipathic C-terminal domain. *Scientific Reports, 6*, 36930.

Janer, A., Prudent, J., Paupe, V., Fahiminiya, S., Majewski, J., Sgarioto, N., et al. (2016). SLC25A46 is required for mitochondrial lipid homeostasis and cristae maintenance and is responsible for Leigh syndrome. *EMBO Molecular Medicine, 8*(9), 1019–1038. https://www.embopress.org/doi/full/10.15252/emmm.201506159.

Jaworski, C. J., Moreira, E., Li, A., Lee, R., & Rodriguez, I. R. (2001). A family of 12 human genes containing oxysterol-binding domains. *Genomics, 78*, 185–196.

Jennings, J. J., et al. (2006). Mitochondrial aberrations in mucolipidosis Type IV. *The Journal of Biological Chemistry, 281*, 39041–39050.

Jiang, B., Brown, J. L., Sheraton, J., Fortin, N., & Bussey, H. (1994). A new family of yeast genes implicated in ergosterol synthesis is related to the human oxysterol binding protein. *Yeast (Chichester, England), 10*, 341–353.

Jiang, H., & Vogt, P. K. (2008). Constitutively active Rheb induces oncogenic transformation. *Oncogene, 27*, 5729–5740.

Johansson, M., Lehto, M., Tanhuanpää, K., Cover, T. L., & Olkkonen, V. M. (2005). The oxysterol-binding protein homologue ORP1L interacts with Rab7 and alters functional properties of late endocytic compartments. *Molecular Biology of the Cell, 16*(12), 5480–5492. https://www.molbiolcell.org/doi/10.1091/mbc.e05-03-0189?url_ver=Z39.88-2003&rfr_id=ori:rid:crossref.org&rfr_dat=cr_pub%20%200pubmed.

Jung, J., Genau, H. M., & Behrends, C. (2015). Amino acid-dependent mTORC1 regulation by the lysosomal membrane protein SLC38A9. *Molecular and Cellular Biology, 35*, 2479–2494.

Kandutsch, A. A., & Chen, H. W. (1977). Consequences of blocked sterol synthesis in cultured cells. DNA synthesis and membrane composition. *The Journal of Biological Chemistry, 252*, 409–415.

Kandutsch, A. A., & Thompson, E. B. (1980). Cytosolic proteins that bind oxygenated sterols. Cellular distribution, specificity, and some properties. *The Journal of Biological Chemistry, 255*, 10813–10821.

Krick, R., Henke, S., Tolstrup, J., & Thumm, M. (2008). Dissecting the localization and function of Atg18, Atg21 and Ygr223c. *Autophagy, 4*, 896–910.

Kumar, N., et al. (2018). VPS13A and VPS13C are lipid transport proteins differentially localized at ER contact sites. *The Journal of Cell Biology, 217*, 3625–3639.

Kvam, E., & Goldfarb, D. S. (2006). Nucleus-vacuole junctions in yeast: Anatomy of a membrane contact site. *Biochemical Society Transactions, 34*, 340–342.

Kvam, E., Gable, K., Dunn, T. M., & Goldfarb, D. S. (2005). Targeting of Tsc13p to nucleus-vacuole junctions: A role for very-long-chain fatty acids in the biogenesis of microautophagic vesicles. *Molecular Biology of the Cell, 16*, 3987–3998.

Laitinen, S., Olkkonen, V. M., Ehnholm, C., & Ikonen, E. (1999). Family of human oxysterol binding protein (OSBP) homologues: A novel member implicated in brain sterol metabolism. *Journal of Lipid Research, 40*, 2204–2211.

Lauzier, A., et al. (2022). Snazarus and its human ortholog SNX25 modulate autophagic flux. *Journal of Cell Science, 135*, jcs258733.

Lehto, M., et al. (2008). The R-Ras interaction partner ORP3 regulates cell adhesion. *Journal of Cell Science, 121*, 695–705.

Lemmon, M. A. (2008). Membrane recognition by phospholipid-binding domains. *Nature Reviews. Molecular Cell Biology, 9*, 99–111.

Lemmon, M. A., Ferguson, K. M., & Abrams, C. S. (2002). Pleckstrin homology domains and the cytoskeleton. *FEBS Letters, 513*, 71–76.

Leonzino, M., Reinisch, K. M., & De Camilli, P. (2021). Insights into VPS13 properties and function reveal a new mechanism of eukaryotic lipid transport. *Biochimica et Biophysica Acta (BBA) – Molecular and Cell Biology of Lipids, 1866*, 159003.

Levine, T. P., & Munro, S. (2001). Dual targeting of Osh1p, a yeast homologue of oxysterol-binding protein, to both the Golgi and the nucleus-vacuole junction. *Molecular Biology of the Cell, 12*, 1633–1644.

Li, J., et al. (2014). SNX13 reduction mediates heart failure through degradative sorting of apoptosis repressor with caspase recruitment domain. *Nature Communications, 5*, 5177.

Liao, P.-C., et al. (2022). Touch and go: Membrane contact sites between lipid droplets and other organelles. *Frontiers in Cell and Developmental Biology, 10*.

Lim, C.-Y., et al. (2019). ER-lysosome contacts enable cholesterol sensing by mTORC1 and drive aberrant growth signalling in Niemann-Pick type C. *Nature Cell Biology, 21*, 1206–1218.

Lu, A., et al. (2022). CRISPR screens for lipid regulators reveal a role for ER-bound SNX13 in lysosomal cholesterol export. *Journal of Cell Biology, 221*, e202105060.

Manifava, M., et al. (2016). Dynamics of mTORC1 activation in response to amino acids. *eLife, 5*, e19960.

Marques, A. R. A., & Saftig, P. (2019). Lysosomal storage disorders – Challenges, concepts and avenues for therapy: Beyond rare diseases. *Journal of Cell Science, 132*, jcs221739.

McCauliff, L. A., et al. (2019). Intracellular cholesterol trafficking is dependent upon NPC2 interaction with lysobisphosphatidic acid. *Elife, 8*, e50832.

Medoh, U. N., Chen, J. Y., & Abu-Remaileh, M. (2022). Lessons from metabolic perturbations in lysosomal storage disorders for neurodegeneration. *Current Opinion in Systems Biology, 29*, 100408.

Melia, T. J., & Reinisch, K. M. (2022). A possible role for VPS13-family proteins in bulk lipid transfer, membrane expansion and organelle biogenesis. *Journal of Cell Science, 135*. jcs259357.

Millen, J. I., Pierson, J., Kvam, E., Olsen, L. J., & Goldfarb, D. S. (2008). The luminal N-terminus of yeast Nvj1 is an inner nuclear membrane anchor. *Traffic (Copenhagen, Denmark), 9*, 1653–1664.

Moskvina, E., Schüller, C., Maurer, C. T., Mager, W. H., & Ruis, H. (1998). A search in the genome of *Saccharomyces cerevisiae* for genes regulated via stress response elements. *Yeast (Chichester, England), 14*, 1041–1050.

Nguyen, T. B., & Olzmann, J. A. (2019). Getting a handle on lipid droplets: Insights into ER–lipid droplet tethering. *The Journal of Cell Biology, 218*, 1089–1091.

Obaseki, E., Adebayo, D., Bandyopadhyay, S., & Hariri, H. (2024). Lipid droplets and fatty acid-induced lipotoxicity: In a nutshell. *FEBS Letters*. https://doi.org/10.1002/1873-3468.14808.

Pan, X., et al. (2000). Nucleus-vacuole junctions in *Saccharomyces cerevisiae* are formed through the direct interaction of Vac8p with Nvj1p. *Molecular Biology of the Cell, 11*, 2445–2457.

Parmar, N., & Tamanoi, F. (2010). *Chapter 3 – Rheb G-proteins and the activation of mTORC1. The enzymes, vol. 27*, Academic Press, 39–56.

Paul, B., et al. (2022). Structural predictions of the SNX-RGS proteins suggest they belong to a new class of lipid transfer proteins. *Frontiers in Cell and Developmental Biology, 10*.

Peng, W., Schröder, L. F., Song, P., Wong, Y. C., & Krainc, D. (2023). Parkin regulates amino acid homeostasis at mitochondria-lysosome (M/L) contact sites in Parkinson's disease. *Science Advances, 9*, eadh3347.

Peng, W., Wong, Y. C., & Krainc, D. (2020). Mitochondria-lysosome contacts regulate mitochondrial Ca2+ dynamics via lysosomal TRPML1. *Proceedings of the National Academy of Sciences of the United States of America, 117*, 19266–19275.

Pijuan, J., et al. (2022). Mitochondrial dynamics and mitochondria-lysosome contacts in neurogenetic diseases. *Frontiers in Neuroscience, 16*.

Radulovic, M., et al. (2022). Cholesterol transfer via endoplasmic reticulum contacts mediates lysosome damage repair. *The EMBO Journal, 41*, e112677.

Ratto, E., et al. (2022). Direct control of lysosomal catabolic activity by mTORC1 through regulation of V-ATPase assembly. *Nature Communications, 13*, 4848.

Roberts, P., et al. (2003). Piecemeal microautophagy of nucleus in *Saccharomyces cerevisiae*. *Molecular Biology of the Cell, 14*, 129–141.

Rogers, S., Hariri, H., Wood, N. E., Speer, N. O., & Henne, W. M. (2021). Glucose restriction drives spatial reorganization of mevalonate metabolism. *eLife, 10*, e62591.

Rudnik, S., & Damme, M. (2021). The lysosomal membrane—Export of metabolites and beyond. *The FEBS Journal, 288*, 4168–4182.

Sabatini, D. D., & Adesnik, M. (2013). Christian de Duve: Explorer of the cell who discovered new organelles by using a centrifuge. *Proceedings of the National Academy of Sciences of the United States of America, 110*, 13234–13235.

Saric, A., et al. (2021). SNX19 restricts endolysosome motility through contacts with the endoplasmic reticulum. *Nature Communications, 12*, 4552.

Schmiege, P., Fine, M., Blobel, G., & Li, X. (2017). Human TRPML1 channel structures in open and closed conformations. *Nature, 550*, 366–370.

Schott, M. B., et al. (2024). Ethanol disrupts hepatocellular lipophagy by altering Rab5-centric LD-lysosome trafficking. *Hepatology Communications, 8*, e0446.

Schröder, L. F., et al. (2024). VPS13C regulates phospho-Rab10-mediated lysosomal function in human dopaminergic neurons. *The Journal of Cell Biology, 223*, e202304042.

Schuettpelz, J., Janer, A., Antonicka, H., & Shoubridge, E. A. (2023). The role of the mitochondrial outer membrane protein SLC25A46 in mitochondrial fission and fusion. *Life Science Alliance, 6*, e202301914.

Schulze, R. J., et al. (2020). Direct lysosome-based autophagy of lipid droplets in hepatocytes. *Proceedings of the National Academy of Sciences of the United States of America, 117*, 32443–32452.

Schwake, M., Schröder, B., & Saftig, P. (2013). Lysosomal membrane proteins and their central role in physiology. *Traffic (Copenhagen, Denmark), 14*, 739–748.

Scotto Rosato, A., et al. (2019). TRPML1 links lysosomal calcium to autophagosome biogenesis through the activation of the CaMKKβ/VPS34 pathway. *Nature Communications, 10*, 5630.

Shai, N., Schuldiner, M., & Zalckvar, E. (2016). No peroxisome is an island—Peroxisome contact sites. *Biochimica et Biophysica Acta (BBA) – Molecular Cell Research, 1863*, 1061–1069.

Short, B. (2015). Mdm1 helps the ER and vacuole stay in touch. *The Journal of Cell Biology, 210*, 522.

Sobajima, T., et al. (2018). The Rab11-binding protein RELCH/KIAA1468 controls intracellular cholesterol distribution. *Journal of Cell Biology, 217*, 1777–1796.

Starling, G. P., et al. (2016). Folliculin directs the formation of a Rab34–RILP complex to control the nutrient-dependent dynamic distribution of lysosomes. *EMBO Reports, 17*, 823–841.

Su, K., et al. (2017). Structure of the PX domain of SNX25 reveals a novel phospholipid recognition model by dimerization in the PX domain. *FEBS Letters, 591*, 2011–2018.

Suh, J. M., et al. (2008). An RGS-containing sorting nexin controls Drosophila lifespan. *PLoS One, 3*, e2152.

Taylor, F. R., Saucier, S. E., Shown, E. P., Parish, E. J., & Kandutsch, A. A. (1984). Correlation between oxysterol binding to a cytosolic binding protein and potency in the repression of hydroxymethylglutaryl coenzyme A reductase. *Journal of Biological Chemistry, 259*, 12382–12387.

Thomas, A. C., et al. (2014a). Mutations in SNX14 cause a distinctive autosomal-recessive cerebellar ataxia and intellectual disability syndrome. *American Journal of Human Genetics, 95*, 611–621.

Thomas, J. D., et al. (2014b). Rab1A is an mTORC1 activator and a colorectal oncogene. *Cancer Cell, 26*, 754–769.

Tong, J., Manik, M. K., & Im, Y. J. (2018). Structural basis of sterol recognition and nonvesicular transport by lipid transfer proteins anchored at membrane contact sites. *Proceedings of the National Academy of Sciences of the United States of America, 115*, E856–E865.

Trivedi, P. C., Bartlett, J. J., & Pulinilkunnil, T. (2020). Lysosomal biology and function: Modern view of cellular debris bin. *Cells, 9*, 1131.

Ugrankar, R., et al. (2019). *Drosophila* Snazarus regulates a lipid droplet population at plasma membrane-droplet contacts in adipocytes. *Developmental Cell, 50*, 557–572.e5.

Van der Kant, R., et al. (2013). Late endosomal transport and tethering are coupled processes controlled by RILP and the cholesterol sensor ORP1L. *Journal of Cell Science, 126*, 3462–3474.

Voeltz, G. K., Sawyer, E. M., Hajnóczky, G., & Prinz, W. A. (2024). Making the connection: How membrane contact sites have changed our view of organelle biology. *Cell, 187*, 257–270.

Wijdeven, R. H., Janssen, H., Nahidiazar, L., Janssen, L., Jalink, K., Berlin, I., et al. (2016). Cholesterol and ORP1L-mediated ER contact sites control autophagosome transport and fusion with the endocytic pathway. *Nature Communications, 7*, 11808. https://www.nature.com/articles/ncomms11808.

Wong, Y. C., Ysselstein, D., & Krainc, D. (2018). Mitochondria–lysosome contacts regulate mitochondrial fission via RAB7 GTP hydrolysis. *Nature, 554*, 382–386.

Xu, H., & Ren, D. (2015). Lysosomal physiology. *Annual Review of Physiology, 77*, 57–80.

Yamano, K., Fogel, A. I., Wang, C., van der Bliek, A. M., & Youle, R. J. (2014). Mitochondrial Rab GAPs govern autophagosome biogenesis during mitophagy. *Elife, 3*, e01612.

Yang, H., Jiang, X., Li, B., Yang, H. J., Miller, M., Yang, A., et al. (2017). Mechanisms of mTORC1 activation by RHEB and inhibition by PRAS40. *Nature, 552*, 368–373. https://www.nature.com/articles/nature25023.

Yeo, H. K., et al. (2021). Phospholipid transfer function of PTPIP51 at mitochondria-associated ER membranes. *EMBO Reports, 22*, e51323.

Zhang, J., et al. (2013). A tuberous sclerosis complex signalling node at the peroxisome regulates mTORC1 and autophagy in response to ROS. *Nature Cell Biology, 15*, 1186–1196.

Zhao, K., & Ridgway, N. D. (2017). Oxysterol-binding protein-related protein 1L regulates cholesterol egress from the endo-lysosomal system. *Cell Reports, 19*, 1807–1818.

Printed in the United States
by Baker & Taylor Publisher Services